D0072204

ENERGY AND
THE ENVIRONMENT

Natural Gas
and Hydrogen

ENERGY AND THE ENVIRONMENT

Natural Gas and Hydrogen

JOHN TABAK, Ph.D.

An imprint of Infobase Publishing

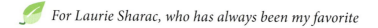 *For Laurie Sharac, who has always been my favorite*

NATURAL GAS AND HYDROGEN

Copyright © 2009 by John Tabak, Ph.D.

Facts On File, Inc.
An imprint of Infobase Publishing
132 West 31st Street
New York NY 10001

Library of Congress Cataloging-in-Publication Data

Tabak, John.
  Natural gas and hydrogen / John Tabak.
      p. cm. — (Energy and the environment)
  Includes bibliographical references and index.
  ISBN-13: 978-0-8160-7084-8
  ISBN-10: 0-8160-7084-9
  1. Natural gas. 2. Gas as fuel. 3. Hydrogen. 4. Hydrogen as fuel. 5. Hydrogen as fuel.
  I. Title.

  TP350.T23 2009
  333.8′23—dc22   2008026072

Facts On File books are available at special discounts when purchased in bulk quantities for businesses, associations, institutions, or sales promotions. Please call our Special Sales Department in New York at (212) 967-8800 or (800) 322-8755.

You can find Facts On File on the World Wide Web at http://www.factsonfile.com

Text design by Erik Lindstrom
Illustrations by Sholto Ainslie
Photo research by Elizabeth H. Oakes

Printed in the United States of America

Bang Hermitage 10 9 8 7 6 5 4 3

This book is printed on acid-free paper.

Contents

Preface

Nations around the world already require staggering amounts of energy for use in the transportation, manufacturing, heating and cooling, and electricity sectors, and energy requirements continue to increase as more people adopt more energy-intensive lifestyles. Meeting this ever-growing demand in a way that minimizes environmental disruption is one of the central problems of the 21st century. Proposed solutions are complex and fraught with unintended consequences.

The six-volume Energy and the Environment set is intended to provide an accessible and comprehensive examination of the history, technology, economics, science, and environmental and social implications, including issues of environmental justice, associated with the acquisition of energy and the production of power. Each volume describes one or more sources of energy and the technology needed to convert it to useful working energy. Considerable emphasis is placed on the science on which the technology is based, the limitations of each technology, the environmental implications of its use, questions of availability and cost, and the way that government policies and energy markets interact. All of these issues are essential to understanding energy. Each volume also includes an interview with a prominent person in the field addressed. Interview topics range from the scientific to the highly personal, and reveal additional and sometimes surprising facts and perspectives.

Nuclear Energy discusses the physics and technology of energy production, reactor design, nuclear safety, the relationship between commercial nuclear power and nuclear proliferation, and attempts by the United States to resolve the problem of nuclear waste disposal. It concludes by contrasting the nuclear policies of Germany, the United States, and France. Harold Denton, former director of the Office of Nuclear Reactor Regulation at the U.S. Nuclear Regulatory Commission, is interviewed about the commercial nuclear industry in the United States.

Biofuels describes the main fuels and the methods by which they are produced as well as their uses in the transportation and electricity-production sectors. It also describes the implications of large-scale biofuel use on the environment and on the economy with special consideration given to its effects on the price of food. The small-scale use of biofuels—for example, biofuel use as a form of recycling—are described in some detail, and the volume concludes with a discussion of some of the effects that government policies have had on the development of biofuel markets. This volume contains an interview with economist Dr. Amani Elobeid, a widely respected expert on ethanol, food security, trade policy, and the international sugar markets. She shares her thoughts on ethanol markets and their effects on the price of food.

Coal and Oil describes the history of these sources of energy. The technology of coal and oil—that is, the mining of coal and the drilling for oil as well as the processing of coal and the refining of oil—are discussed in detail, as are the methods by which these primary energy sources are converted into useful working energy. Special attention is given to the environmental effects, both local and global, associated with their use and the relationships that have developed between governments and industries in the coal and oil sectors. The volume contains an interview with Charlene Marshall, member of the West Virginia House of Delegates and vice chair of the Select Committee on Mine Safety, about some of the personal costs of the nation's dependence on coal.

Natural Gas and Hydrogen describes the technology and scale of the infrastructure that have evolved to produce, transport, and consume natural gas. It emphasizes the business of natural gas production and the energy futures markets that have evolved as vehicles for both speculation and risk management. Hydrogen, a fuel that continues to attract a great deal of attention and research, is also described. The book focuses on possible advantages to the adoption of hydrogen as well as the barriers that have so far prevented large-scale fuel-switching. This volume contains an interview with Dr. Ray Boswell of the U.S. Department of Energy's National Energy Technology Laboratory about his work in identifying and characterizing methane hydrate reserves, certainly one of the most promising fields of energy research today.

Wind and Water describes conventional hydropower, now-conventional wind power, and newer technologies (with less certain futures) that are being introduced to harness the power of ocean currents, ocean waves, and the temperature difference between the upper and lower layers of the ocean. The strengths and limitations of each technology are discussed at some length, as are mathematical models that describe the maximum amount of energy that can be harnessed by such devices. This volume contains an interview with Dr. Stan Bull, former associate director for science and technology at the National Renewable Energy Laboratory, in which he shares his views about how scientific research is (or should be) managed, nurtured, and evaluated.

Solar and Geothermal Energy describes two of the least objectionable means by which electricity is generated today. In addition to describing the nature of solar and geothermal energy and the

processes by which these sources of energy can be harnessed, it details how they are used in practice to supply electricity to the power
markets. In particular, the reader is introduced to the difference
between base load and peak power and some of the practical differences between harnessing an intermittent energy source (solar) and
a source that can work virtually continuously (geothermal). Each
section also contains a discussion of some of the ways that govern-
mental policies have been used to encourage the growth of these
sectors of the energy markets. The interview in this volume is with
John Farison, director of Process Engineering for Calpine Corporation at the Geysers Geothermal Field, one of the world's largest and
most productive geothermal facilities, about some of the challenges
of running and maintaining output at the facility.

Energy and the Environment is an accessible and comprehensive
introduction to the science, economics, technology, and environmental and societal consequences of large-scale energy production
and consumption. Photographs, graphs, and line art accompany the
text. While each volume stands alone, the set can also be used as a
reference work in a multidisciplinary science curriculum.

Acknowledgments

The author thanks Ray Boswell, for his thoughtful answers and the generous way that he shared his time and insights, and Sandra Schlenker and the West Bloomfield Historical Society, for their help in obtaining information about the earliest days of the natural gas industry in the United States. Finally, a note of appreciation for the assistance of Ken McDonnell, Senior Media Relations Specialist with ISO New England; Elizabeth Oakes, for her fine photo research; and Frank K. Darmstadt, executive editor, Facts On File, for the vote of confidence.

Introduction

Whether solid, liquid, or gaseous, the phase of a fuel has important implications for how it can be used. *Natural Gas and Hydrogen* is concerned with the two most important gaseous fuels. Natural gas derives much of its importance from the residential and commercial heating markets and the fact that a great deal of electricity is generated by natural gas–fired power plants. (Approximately 20 percent of the electricity in the United States is obtained from natural gas–fired units.) Hydrogen derives its importance from its promise: It may one day be competitive with oil as a transportation fuel.

The first section of this book is concerned with natural gas and related fuels, including *coal gas,* a manufactured gas, the use of which preceded natural gas. Gaseous fuels are best transported by pipeline. The dependence of the natural gas industry on pipelines has had a large effect on the ways that the industry has developed.

From an early date, the federal government perceived natural gas pipelines as "natural monopolies," and sought to create regulatory schemes that would protect both consumers and producers of natural gas from the potentially monopolistic practices of the pipeline operators. These efforts had profound effects on the natural gas business. Early regulatory efforts and the early markets that developed under these regulatory schemes are described in chapter 1, as is the history of the now-defunct coal gas industry.

The geology of natural gas—how it is formed, where it is found, and some of the methods used to obtain it—is covered in chapter 2, as are various related synthetic gases. *Methane hydrates,* the largest potential source of methane on the planet, are also described. (Methane hydrates currently contribute almost nothing to the world's energy supply, but they could potentially contribute more to the world's supply of energy than all conventional fossil fuels put together. For more information, see the interview with Dr. Ray Boswell in chapter 6.)

Some aspects of the technology used to transport and store natural gas are described in chapter 3. It is difficult to appreciate how the natural gas industry has developed (and is developing) without a clear understanding of these vital technologies.

Natural gas, because it is a fossil fuel, has much in common with other fossil fuels in terms of its effects on the environment. Chapter 4 describes the phenomenon of combustion and the way that the chemical energy contained in natural gas is converted into electricity. Central to these ideas is the concept of efficiency, which is the ratio of the amount of electrical energy produced by a power plant to the amount of thermal energy supplied. Finally, the effects on the environment of burning natural gas are discussed.

The current natural gas markets in the United States, which are products of a relatively new and still evolving federal regulatory scheme, are described in chapter 5. The restructured marketplace has changed how natural gas is priced and to some extent how it is used. Natural gas consumers are now more directly affected by the volatile price of natural gas, the expression of which is made possible by the new markets. Chapter 6 describes some of the ways that producers and large-scale consumers of natural gas attempt to protect themselves from fluctuating gas prices, which is one aspect of the natural gas futures markets. Both chapters 5 and 6 are about the business end of the natural gas business, an important topic, because, in the end, the natural gas business is about profit first and natural gas second. Natural gas production is the vehicle by which these companies earn their profits. Energy futures markets are an important way that profits are protected and sometimes generated. Some connections between government regulations and the "free" market are also discussed.

The second part of the book is about hydrogen. The *hydrogen economy,* a concept and a phrase that date back to the 1970s, arose out of the so-far unrealized hope that hydrogen will replace petroleum as a transportation fuel. A generation of researchers has already spent its professional life attempting to create a hydrogen economy, but the simply-stated idea of using hydrogen instead of gasoline in automobiles has turned out to be remarkably difficult to bring to fruition.

Chapter 7 describes the main barriers that must be overcome if a hydrogen economy is to take shape. These are the large-scale production of hydrogen, the problems of hydrogen transport and storage, and the problem of converting the chemical energy carried by hydrogen into work. Special attention is given to the hydrogen-powered *fuel cell,* a device that converts the chemical energy of hydrogen directly into electricity.

Chapter 8 describes why, despite the very substantial difficulties involved in using hydrogen, the hydrogen economy remains such an attractive concept to so many and how some researchers believe the United States can transition to hydrogen from petroleum. Finally, chapter 9 describes some of the ways that the federal government has attempted (and is attempting) to create a hydrogen economy.

Federal efforts are extremely important and extremely interesting because of their scope. Large-scale, federally funded research programs are attempting to create a new global economy, one based on hydrogen rather than petroleum, and if successful this research will have profound consequences for every creature on the planet.

Energy—finding it, transporting it, storing it, and harnessing it—may be the most important issue of the 21st century. Natural gas will remain an important fuel for decades, perhaps much longer if a way is found to exploit methane hydrates. Some believe that hydrogen will be as important to the latter decades of the 21st century as oil was to the latter days of the 20th. Understanding the value of these gaseous fuels and what their use means (or might mean) to the economy and the environment involves appreciating a complex interplay of economics, science, engineering, legislation, and regulation. These are fascinating and important topics.

PART I

Natural Gas

Early Gas Technologies and Policies

There are three major fossil fuels—coal, oil, and natural gas—and of the three, natural gas was the last to be targeted for large-scale production. In fact, for many decades, energy companies preferred to manufacture "coal gas," a toxic gas that, when burned, released much less energy than the same volume of natural gas. Their reasons for preferring coal gas to natural gas illustrate many of the difficulties involved in the early commercial utilization of natural gas. This chapter begins, therefore, with the history of coal gas, also called town gas, even though coal gas is an energy source that is very different from natural gas.

Early attempts to exploit natural gas are, to a modern reader, sometimes humorous and sometimes shocking. Many of the earliest attempts to use natural gas ended in explosions. Others succumbed to the most primitive attempts at sabotage. This chapter

Coal gas plant, Los Angeles, California. These plants were once found in cities around the world. *(Dr. Allen W. Hatheway, Rolla, Missouri)*

describes both, because before the natural gas industry could flourish, engineers would have to find solutions to these types of technological challenges.

Just as important as technology, however, was the unregulated nature of the early gas business. Throughout most of the 19th-century, individuals, industry, and government perceived natural gas resources as so huge that they could not be exhausted. "Limitless" was a commonly used word for describing the volumes of natural gas that had been discovered in Pennsylvania and Indiana. This perception caused individuals and companies to waste astonishing amounts of gas. Government regulations, which have always played an important part in the history of natural gas, were introduced too late to affect the development of some of these early gas fields. Without a regulatory framework, there was no rational reason for any individual or company to restrict consumption. In fact, to restrict consumption under these circumstances was to provide competitors with a substantial advantage. This perception of natural gas stocks as unlimited, which is described in the last section of the chapter, had to be overcome before natural gas could be used in a way that was economically sensible.

COAL GAS

For about a century, gas manufactured from coal—not natural gas, which is produced from wells much as oil is—was the fuel of choice for many cities. Coal gas, also known as town gas, was the product of a process pioneered by the British inventor William Murdock (1754–1839). He discovered that by heating coal in a low-oxygen environment, some of the coal could be converted to a flammable gas. (Another early innovator, the French engineer and inventor Philippe Lebon [1767–1804], did essentially the same thing—producing gas by heating wood.) Depending on the details of the process used to create the gas—and depending on the chemical composition of the coal—it was possible to convert up to about 40 percent of the coal into a volatile gas. The resulting gas was cooled and cleaned of a sticky material called coal tar as well as ammonia and the smelly sulfur compounds that were also produced during the process. After further cooling, the clean (or at least cleaner) gas was ready to be burned. Coal gas consisted of a mixture of (poisonous) carbon monoxide, hydrogen, methane, and whatever impurities the cleaning processes failed to remove.

At the time, coal gas had a number of advantages over natural gas. First, coal, which provided the raw material from which coal gas was produced, also provided the energy source that powered the conversion process. Second, coal was readily available. Third, coal, because it is a solid, was relatively easy to transport from the mines where it was produced to the *gasworks* in the cities where it was needed. Fourth, coal was easy to store. In a sense, coal was coal gas in solid form—transported as a solid and stored as a solid. Coal gas was then manufactured on-site in accordance with demand.

By contrast, the pipeline technology needed to transport large volumes of natural gas from the fields where it was produced to the (often) distant cities where it was needed would not be developed until the 20th century. Nor had engineers yet developed a way to safely store large quantities of natural gas. The transport and storage

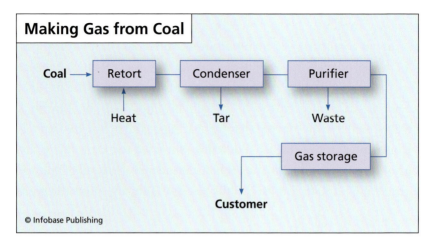

Making Gas from Coal

Coal → Retort → Condenser → Purifier → Gas storage → Customer

Heat   Tar   Waste

© Infobase Publishing

Process diagram for the manufacture of coal gas. Significant amounts of pollution were generated at each step of the process to produce a product (coal gas) that was poisonous.

problems that stymied the development of the early natural industry were, therefore, circumvented by using coal gas. To be sure, the coal gas manufacturing process was dirty. It resulted in a lot of air pollution as well as a number of harmful byproducts that were often tossed on the ground around the gasworks, thereby polluting both the land and the water. But pollution was not a primary concern for 19th-century city dwellers; street illumination was. And coal gas, produced at local gasworks and transported to streetlamps by small-scale systems of pipes, offered an essential service for which there was no alternative.

The first commercial concern to begin supplying coal gas was the London & Westminster Gas Light & Coke Company, which began operation in 1812. (*Coke* was the name given to the solid that remained after the coal gas had been produced. It was—and in a more purified form still is—used in metallurgy.) In the United States, many coal gas companies were established shortly after London & Westminster received its charter. In 1816, a coal gas infrastructure was under construction in Baltimore, Maryland. New York City was next in 1823, and Boston was third in 1828. Pipelines were soon extended from streets into houses as coal gas became the preferred indoor illuminant, replacing whale oil, which was becoming increasingly expensive as unrestricted whaling decimated whale populations. Tallow candles, another common light source of the time, were inexpensive but dirty; they produced a lot of smoke and only a little light. Gaslights were a dramatic improvement over these other sources of illumination. They were brighter, cheaper, and more convenient. Demand soared. By 1859, there were 297 gas companies in the United States, supplying coal gas, principally for illumination, to 4,860,000 customers.

The coal gas companies were generally regulated by the municipalities in which they were located because they were "natural monopolies"—that is, in contrast to candle makers who competed freely and sometimes had geographically overlapping service areas, there was only enough room on any given street for a single set of gas pipes. Once the pipes were in place, therefore, no further competition between coal gas suppliers was possible. To prevent the abuses commonly associated with monopolies, each municipality regulated the gasworks located within its boundaries. In the early days of the industry, a large city would often be serviced by several coal gas companies. Each company had its own service area, and each service area would be divided into smaller regions, each with its own gasworks. Local regulatory agencies were all that were needed to regulate these local monopolies. Even if a coal gas company wanted to service multiple cities in multiple states, it could not accomplish this from a single facility because the technology needed to transport large volumes of gas across large distances had not yet been invented.

During the latter decades of the 19th century, the many small companies that originally comprised the industry began to consolidate into a smaller number of much larger companies, but the coal

(continues on page 10)

The End of Coal Gas

Today, energy crises of various kinds are the subjects of frequent news reports, as are reports on the necessity of replacing one source of energy with another. Often little attention is paid to the difficulties involved. It is interesting, therefore, to examine the fate of coal gas, which was once the only commercially viable gaseous fuel.

The first use of coal gas was as a source of illumination, first for streetlights and later for indoor lighting. In 1879, the American inventor Thomas Alva Edison (1847–1931) developed a prototype of an incandescent light bulb, and by 1880 Edison's bulbs were illuminating the steamship *Columbia,* the first demonstration of their practical utility. One might think that Edison's invention marked the end of coal gas, or at least the

A period cartoon showing Edison as hero and the coal gas manufacturers as villains. It would, however, be decades before the coal gas industry went into decline. *(Library of Congress. Published in October 23, 1878, in* Puck, *p. 16.)*

end of the era of gaslight illumination, but the first effect of the incandescent light on the coal gas industry was to accelerate gaslight technology. In 1885, the Austrian engineer and chemist Carl Auer, baron von Welsbach (1858–1929) patented the Welsbach light, which enabled gaslights to burn much more brightly using the same amount of fuel as a conventional gas light. Other innovations in gas lighting followed.

In the United States, it took about 40 years for the electric light to displace the gaslight, and the reasons are not hard to appreciate. First, those who wanted electricity had to pay—not just for the lights but for the infrastructure needed to produce and distribute the electricity needed to power the lights. Power stations had to be built, wires strung, individual houses had to be wired, and electric lights had to be purchased. Second, in contrast to electric light technology, which initially was not especially reliable, coal gas technology was reliable, and whatever its shortcomings, an enormous coal gas infrastructure was already in place, providing light and heat to millions of consumers throughout the nation. Consumers had already spent substantial sums of money outfitting their houses with pipes and the gas appliances required to make use of the fuel. Having spent this money, many consumers were reluctant to stop using that for which they had already paid.

But even as the electric light made inroads into the illumination business, coal gas producers continued to flourish because they found new markets for their product. Cooking and heating became important applications. One might think that natural gas, which has a higher energy content than coal gas, was better suited to these markets, but coal gas could compete, because (again) the coal gas infrastructure was already in place.

Nor was the *higher heating value* of natural gas its only advantage over coal gas. The process of manufacturing coal gas produced large amounts of pollution, some of it quite dangerous, and this pollution was produced within the city limits. By contrast, natural gas burned far more cleanly than coal gas, and it was piped in from distant fields, so production was

(continues)

(continued)

completely out of sight. From the consumer's point of view, therefore, natural gas was practically pollution free. Despite these advantages, coal gas's position as the first fuel in place gave it an advantage over natural gas and electricity that lasted for decades.

It was not until World War II, when a large interstate pipeline system was brought online, that coal gas usage began to decline sharply. The last gasworks in the United States closed in the 1960s, demonstrating just how long it can take to replace one energy source with others, even when the replacements are demonstrably superior.

(continued from page 7)

gas business remained local in nature, and local regulatory bodies remained sufficient to control the industry. The way that coal gas was used also began to change in response to changing market conditions. In particular, with the advent of Thomas Edison's incandescent light, the demand for gaslights began to fade and the coal gas industry began to market coal gas for use as a heating and cooking fuel. So even as the demand for gas lighting slowly diminished—and demand diminished surprisingly slowly—the coal gas industry continued to flourish. It remained a major industry for the first few decades of the 20th century.

EARLY ATTEMPTS TO USE NATURAL GAS

The first successful attempt to use natural gas in modern historical times occurred in Fredonia, New York. As early as 1820, residents of the area had known that natural gas flowed spontaneously from the ground in various places, and they sometimes intentionally set it alight. William Hart, a gunsmith and local entrepreneur, recognized the potential of the discovery and drilled a well. He struck gas at a depth of 27 feet (8 m). This gas was piped to a local inn. It was not until 1858 that a group of local investors capitalized on Hart's discovery and formed the first natural gas company, the Fredonia Gas Light Company.

The small size of Hart's effort and the small size of the Fredonia Gas Light enterprise meant that any profits generated by these projects were small, but the simple technology of the time permitted nothing larger. The history of one of the first large-scale projects, the gas pipeline built by the Bloomfield & Rochester Natural Gas Light Company, illustrates just how primitive engineering practices of the time were. In 1865, gas was struck in West Bloomfield, New York, by drillers in search of oil. There was quite a bit of gas. When the well struck gas at 480 feet (146 m), the pressure of the gas was sufficient to drive a 700-pound (300-kg) chisel and accompanying weights and 408 feet (124 m) of two-inch (5-cm) steel cable back up and out of the well like a cork from a bottle. But when no oil followed the rush of gas, the project, considered a failure, was abandoned. The farmer on whose land the well was located eventually piped the gas from the well to his farmhouse, where he used it for light and heat. (In 1867, the well was ignited and produced a 30-foot [9-m] flame, which was allowed to burn unimpeded and continually. In fact, for the next few years thereafter dances were held each Thursday night by the light of the well fire. They attracted many people.)

In 1870, a group of investors bought the well and surrounding land. Their company, the Bloomfield & Rochester Natural Gas Light Company, was formed to transport the gas from West Bloomfield to Rochester, New York, via pipeline, where it would be sold to the Rochester Gaslight Company, a coal gas company.

The investors quickly raised $100,000 through the sale of stock, and acquired the rights to lay a pipeline along a 30-mile (48-km) corridor connecting West Bloomfield with Rochester. The pipeline

Gas well fire in Coffeyville, Kansas, 1894 *(Allen W. Hatheway)*

was made of wood with a 12.5-inch (32-cm) outer diameter and an eight-inch (20-cm) inner diameter. Sealed inside and out with tar, the segments were carefully joined with metal bands and sealed with cloth soaked in tar. It was the best pipeline technology available, and the investors believed that the pressure of the well would drive the gas all the way to Rochester, producing several hundred thousand cubic feet of gas per day. But when the well was finally completed in 1872, only a tiny fraction of the expected amount issued from the far end of the pipeline. Moreover, Rochester Gaslight almost immediately refused to take delivery of the gas because the mixture of coal gas and natural gas burned poorly—Rochester Gaslight blamed it on the quality of the natural gas—and the Bloomfield & Rochester Natural Gas Light Company went bankrupt shortly thereafter.

An 1894 article in the *Rochester Democrat & Chronicle* newspaper offered a belated explanation for the low rate of flow through the Bloomfield and Rochester pipeline. Several years after the bankruptcy, a farmer (through whose land the pipeline ran) reported striking the old wooden pipeline while digging a ditch. He pulled the pipe out of the ground and reported finding a pair of overalls stuffed inside the pipe, an apparently successful attempt at "pants sabotage." Although the idea to use natural gas as a fuel source was a sound one, the technology needed to transport natural gas from one place to the next was simply not adequate.

But the barriers to effectively exploiting natural gas were more than technical. In fact, late-19th-century attempts to exploit natural gas are, to a modern reader, almost shocking because of the tremendous amounts of waste involved. There were three reasons for the waste. First, there was not a clear understanding of the physics of natural gas deposits. If gas is withdrawn improperly from a deposit, it is possible, on the one hand, to leave huge amounts of gas stranded beneath the ground, while on the other hand, to be left with nothing but unproductive (dry) wells. Maximizing the rate of return from a gas field depends on a better understanding of physics and geology than existed at the time. Second, the technology used to extract and burn the gas was primitive by today's standards. But the most striking aspect of 19th-century attempts to use natural gas was the willingness to deliberately waste it. Because they believed that the supply was limitless, people of the time saw little need to use the resource economically. The exploitation of the Trenton Field, which covered approximately 5,000 square miles (13,000 sq. km) of Indiana, illustrates this attitude.

Discovered in the 1870s, what came to be known as the Trenton Field was initially ignored by drillers in search of oil. But as the magnitude of the resource became apparent, and as the value of natural gas as a heat source became better appreciated, virtually every hamlet in the area scraped together enough investors to sink a well or two. Others came from outside the area to drill. During the height of the drilling frenzy, there were more than 200 companies sinking thousands of wells into the Trenton Field.

Cities and towns alike lured industry into the area, often by promising free gas. Glass and metal industries, which use large quantities of thermal energy, were especially eager to exploit this enormous energy source. And when gas was sold, it was often sold by subscription—that is, each subscriber paid a flat fee per month or per year without regard to how much gas was actually used. One result was the proliferation of flambeaux lights, great flaming natural gas-fired torches that were used to illuminate streets and often left on during the day. Flambeaux lights consumed far more gas than the gaslights favored by those municipalities that depended on metered town gas. Another decorative idea used iron pipes filled with holes and bent into arches. These were connected to a gas supply and left to burn, sometimes for months at a time, thereby creating great arcs of fire.

Initially, natural gas flowed from the Trenton Field at 22 times atmospheric pressure. At the rate gas was extracted, however, that pressure could not be maintained long. In 1890, the field produced approximately 40 billion cubic feet of natural gas. By 1902, the field was producing at five to six times atmospheric pressure. Soon, well after well stopped producing. Companies that depended on the Trenton Field's gas left or went bankrupt. The boom was over. A valuable natural resource had been wasted. For years, this pattern was repeated wherever large reservoirs of natural gas were exploited.

EARLY ATTEMPTS AT REGULATION

The history of regulation of the natural gas markets in the United States is important because regulations have had an important effect on the development of the natural gas industry. Initially, there existed almost no concept of regulation. Everyone behaved as if supplies of natural gas were limitless and, once the cost of drilling a well had been paid, the owners often behaved as if the gas was free. By contrast, this was never the case with coal gas, because coal gas had to be manufactured before it could be used: The more coal gas one used, the more coal one had to transport and process. But if attitudes differed toward the fuels of the two industries, both industries also shared an important characteristic: They were both "natural monopolies." This is easy to see in the coal gas industry, where a single gasworks provided exclusive service to a particular geographical area. In the natural gas business, the monopoly occurs in the transport sector—that is, the pipeline owner can, in theory, exert sole control over the distribution of gas even when there are many producers.

Experience has shown that over time, monopolies, if left unrestrained, tend to decrease the level of service offered their customers while simultaneously increasing their prices. With respect to the coal gas industry, governments recognized the potential hazards to the consumer right from the beginning. Coal gas, whether it was used for illumination, cooking, or heating, was an essential service. But the coal gas company's first responsibility was to generate the maximum possible profit for its investors. Manufacturing and supplying gas were only the means by which this was accomplished, and what was good for the company was not always good for the consumer. Each town or city with a gasworks was, therefore, left with one of two choices: It could regulate the private coal gas supplier and attempt to balance the needs of the private company with the needs of the consumer, or it could establish a municipally owned coal gas system and attempt to run the gasworks itself for the benefit of the municipality. Both approaches were tried at different times and in different places, and each had its successes and failures. What both approaches had in common was that they were local solutions to local problems. Municipal governments were well-suited to the task of finding local solutions, because coal gas systems were, as explained earlier, local systems.

In the natural gas industry, there have always been many companies involved in the production of natural gas. There has never been a monopoly in the production end of the business. But natural gas is rarely produced where it is used. Instead, it must be transported, and the distribution system for natural gas is a natural monopoly for the same reason that a coal gas company is a local monopoly. While there may be competition about which company will build a pipeline network (or a gasworks), once a pipeline network has been built, its existence precludes the construction of a rival network. What distinguishes natural gas pipeline networks from gasworks pipeline networks is that the natural gas networks extend across much larger regions. Even in 1891, the Indiana Natural Gas and Oil Company laid twin natural gas pipelines a distance of 120 miles to connect Chicago with the gas fields to the south. (These pipes were only eight inches [20 cm] in diameter.) Soon pipeline technology improved, and larger pipelines were extended across much greater distances. These geographically larger systems were beyond the reach of the regulatory power of any individual municipality. Consequently, as natural gas distribution networks began to develop, state governments established regulatory bodies that were scaled-up versions of their municipal counterparts.

Throughout the 19th century, pipeline technology continued to improve—by the early 1900s, a 165-mile (264-km) long, 20-inch (51-cm) diameter pipeline was supplying gas to Pittsburgh—and with the development of welded steel pipelines in the 1920s, the industry began to construct even larger interstate distribution systems, systems that were too big for any individual state to effectively regulate.

The federal government's first attempt to regulate the natural gas industry occurred in 1938 with the passage of the Natural Gas Act. The agency charged with implementing the act was the Federal Power Commission, which was established in 1920 and was the predecessor to today's Federal Energy Regulatory Commission. Congress authorized the Federal Power Commission to set "just and reasonable rates" for interstate natural gas sales, which involved the transport of natural gas across state lines via pipeline networks. In addition, the act enabled the Federal Power Commission to regulate the construction of new pipelines, and before any pipeline could be abandoned, the owner had to obtain federal approval. Finally, pipeline companies could pass along the costs of building and operating their pipelines to consumers.

The Natural Gas Act of 1938 had several positive effects. On the one hand, it protected consumers from the excesses often associated with monopolistic business practices. It also stabilized the industry by essentially guaranteeing a profit to those companies that built expensive pipeline projects. The amount of profit earned by the company or companies involved in a pipeline project was regulated by the government, but in an industry accustomed to cycles of boom and bust, the Natural Gas Act of 1938 brought predictability to both consumers and the pipeline companies that supplied them. (The price of gas at the wellhead was not regulated at this time.)

The Natural Gas Act was considerably extended in 1954, when, in the case known as *Phillips Petroleum v. Wisconsin,* the Supreme Court ruled that in addition to interstate pipeline operators, natural gas producers who sold gas that was transported by interstate pipelines were also subject to regulation under the Natural Gas Act. Whatever its merits as law, the Court's decision was poor energy policy. The regulatory burden imposed on the Federal Power Commission proved to be more than it could manage. Because there were many natural gas producers, and the costs of developing natural gas wells are highly dependent on the conditions at the site of the well, establishing reasonable rates for thousands of individual cases was beyond the ability of the Commission. The Commission responded by attempting alternative strategies, such as establishing regional rates, but the task remained more than the agency could successfully execute. The natural gas markets were disrupted. The profits

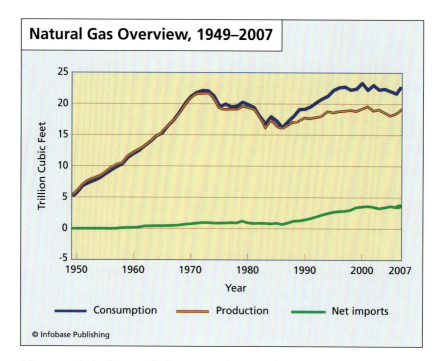

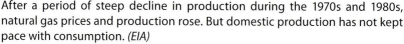

After a period of steep decline in production during the 1970s and 1980s, natural gas prices and production rose. But domestic production has not kept pace with consumption. *(EIA)*

earned by producers due to the sale of gas began to lag behind the costs required for investment in new wells to produce new gas, and production began to falter. Meanwhile, the low prices enjoyed by users of interstate gas, prices that were the result of unreasonable regulation, spurred demand for more gas.

The situation was exacerbated when, during the late 1960s and early 1970s, producers began selling disproportionate amounts of natural gas to markets inside the states where the gas was produced. These *intra*state markets were beyond the reach of federal regulation, and state governments, which were generally more sensitive to the interests of the producers located within their borders, tended to allow the producers to sell gas in-state at higher prices than the federal government would have permitted had they sold their product through the interstate pipeline system. Not surprisingly, markets that depended on interstate pipelines for their supply of natural gas began to experience fuel shortages. In response to this situation, Congress enacted new legislation called the Natural Gas Policy Act of 1978, a complicated piece of legislation that, among other goals, sought over a period of several years to transition the gas markets to a condition in which "market forces" would establish gas prices.

It is important to understand the state of the market prior to the Natural Gas Policy Act of 1978: Those consumers that depended on interstate shipments of gas experienced shortages *because* they constituted an unprofitable market for gas producers. Producers were selling their gas to in-state consumers at rates that they could not have charged if they had sent their gas through the interstate pipeline networks. Compared with out-of-state consumers, consumers inside producing states generally paid more for their gas, but as a consequence of higher prices, their supplies were secure.

The Natural Gas Policy Act of 1978 (NGPA) also granted the Federal Energy Regulatory Commission (FERC) authority over intrastate as well as interstate production. Initially, the NGPA maintained a complicated system of price controls over the production of natural gas. The maximum price that producers could receive depended upon when the well was started, the depth of the well, the date that the well began production, and a number of other factors. This was an extension of the previous system. What was different was that the price ceilings were temporary. They provided stability to the markets while they simultaneously moved the natural gas markets to what the law's authors hoped would be a more sustainable system, one based more on so-called free market principles. This legislation is often described as the first step in creating the restructured natural gas markets that exist today. It was also the first conceptual move away from the regulatory concepts developed during the 19th century by cities for the local coal gas markets. (The current market situation, which evolved from the NGPA, is described in chapter 5.)

One other piece of legislation from the 1970s bears mention: the Power Plant and Industrial Fuel Use Act of 1978. The shortages experienced by consumers living in states that produced little or no natural gas caused Congress to diminish the potential size of the natural gas market in order to bring demand more in line with supply. In particular, Congress restricted the construction of power plants that used natural gas as a fuel. The restriction was more important to the natural gas industry that it was to the electricity industry. The late 1970s and early 1980s was a period of economic stagnation. There was not much demand for new power plants of any kind. In fact, there was a surplus of power capacity. While the NGPA represented a step toward a more modern conception of natural gas markets, the Fuel Use Act of 1978 is sometimes described as a step backward, although the idea still has its supporters. In any case, it was repealed in 1987. During the 20-year period following its repeal (1988–2007), natural gas consumption for the purpose of electrical generation increased from 2.6 trillion cubic feet (73 billion cubic meters) to 6.9 trillion cubic feet (190 billion cubic meters) per year, a situation that has placed further strain on the nation's increasingly tight natural gas supplies.

The Nature of
Natural Gas

The term *natural gas* is applied both to the gas delivered to homes by pipeline and the gas found beneath the surface of the Earth from which the home heating fuel was derived, but chemically the two gases may be quite different. The natural gas at the wellhead is, as a general rule, processed before it is sold. Pipeline operators require that before any gas is transported via their pipelines that it meets certain standards with respect to its chemical composition. This ensures that they can operate their pipelines safely and efficiently. And consumers expect that the gas delivered to homes and businesses will burn predictably in furnaces, water heaters, and other appliances. This can happen only if the chemical composition of the fuel is carefully controlled. One goal of this chapter is to describe a little of the chemistry of natural gas.

Enormous amounts of natural gas exist within a few miles of the Earth's surface. Although estimates vary, most authoritative

Gulf of Mexico offshore drilling platform. High prices for natural gas and oil justify high-cost technology. *(Chad Teer)*

sources agree that more than one thousand trillion standard cubic feet (30,000 billion cubic meters) of recoverable natural gas can be found in the United States alone. There is, therefore, no impending shortage of natural gas. In fact, while estimates of the amount of commercially recoverable natural gas in the United States have fluctuated over the years, there is no clear downward trend in estimates of how much is left. This has long been true despite the fact that large amounts of gas are removed from the ground every day. As new technologies develop, new gas reserves are discovered and deposits of gas that were once thought too difficult or too expensive to develop begin to yield significant quantities of the fuel. The second goal of this chapter is to describe the geological conditions under which natural gas is found and some of the ways that geology affects production.

Once processed, natural gas is almost pure methane, an invisible and odorless gas. There are other sources of pure methane besides natural gas. The greatest potential sources of methane on the planet are deposits of methane hydrates, ice-like solids found deep beneath the sea and under the Arctic tundra. There is far more methane in deposits of methane hydrates than there is in all conventional sources of natural gas. No one has found a way of economically and safely converting methane hydrates into methane gas, but if the technical problems are solved, methane hydrate deposits would have a historic impact on the way that energy is used around the world.

Natural gas can also be manufactured from coal. Chemically, this manufactured gas is methane; it has little in common with the coal gas described in the preceding chapter. This coal-to-methane technology may become more important if the price of natural gas goes high enough to warrant the widespread adoption of the technology. Other types of synthetic gaseous fuels are also being manufactured today. Descriptions of these alternative sources of natural gas and natural gas–like fuels are also to be found within this chapter.

THE ORIGIN OF NATURAL GAS

The gas that flows out of a natural gas well is in the gaseous phase (instead of the liquid or solid phase) because at atmospheric pressure the boiling point of its chemical constituents is lower than the temperature at the top of the well. Methane, the principal chemical constituent of natural gas, has a boiling point of –258.7°F (–161.5°C) at atmospheric pressure. Natural gas deposits also often contain smaller quantities of the gases ethane and propane. At atmospheric pressure ethane boils at a temperature of –127°F (–88.5°C) and propane boils at –43.0°F (–42.2°C). All three gases are examples of hydrocarbons, chemicals whose molecules consist exclusively of collections of hydrogen and carbon atoms. Methane has the simplest structure of the three hydrocarbons just described. A methane molecule consists of a single carbon atom bonded to four hydrogen atoms. Its chemical formula is CH_4, where the letter C is the chemical symbol for carbon, H is the chemical symbol for hydrogen, and the subscript 4 indicates that four hydrogen atoms are bound to one carbon atom. (When there is no subscript, it is understood that only one atom of that type is found in the molecule.) By way of contrast, ethane, whose chemical formula is C_2H_6, consists of two carbon atoms and six hydrogen atoms bound together. The chemical symbol for propane is C_3H_8. These gases are all combustible.

In addition to these three gaseous hydrocarbons, there may also be significant quantities of other chemicals in natural gas deposits, materials that are produced along with the natural gas. These may include oil and water (which are often found in conjunction with natural gas), carbon dioxide, nitrogen (which is the main chemical constituent of air), and small amounts of helium.

The chemical composition of natural gas as it is found within the Earth is a function of the process by which it was formed. It is a complex process that acts on the remains of plants and animals entombed deep within the Earth over extremely long periods of time. The process begins when the remains of these creatures become mixed with and covered by fine sediments. Over time new sediments may accumulate on top of the old, and when this happens, the pressure exerted on the lower layers increases, and the temperature increases as well. As a consequence, the organic matter, now buried deep beneath the surface, is slowly transformed. Given enough time, heat, and pressure, some of this material will be transformed into oil and natural gas. (Where there is oil, there is usually natural gas.) If the organic material is buried deep enough, deeper than about 16,000 feet (5,000 m), only methane will form. Finding and exploiting these deposits of petroleum—a word that denotes both oil and natural gas—forms the basis for one of the biggest and most important industries in the world.

For a long time, most of the emphasis on discovering petroleum deposits was on finding and drilling into geological formations called anticlines, formations that experience has shown are sometimes rich with gas and oil. The gas and oil deposits found in anticlines are thought to have formed in the following way: First, gas and oil form in what are called *source rocks,* the geological formations in which the organic material was first deposited. Source rocks generally have a very fine consistency and whatever petroleum is present is dispersed throughout the formation in such a way that there is not much gas or oil per unit volume. Next, the petroleum, usually mixed with water, begins to migrate as the result of a pressure difference between the source rocks and what are called *reservoir rocks.* Reservoir rocks are more porous than the source rocks. They hold more gas and liquid per unit volume than the source rocks, and petroleum and water are able to move more freely through them. In fact, once in motion—a very slow motion—the petroleum-water mixture will continue to migrate until something causes the mixture to stop. In order for the mixture to accumulate within the reservoir rocks, some sort of petroleum trap must exist.

The trapping function is often performed by an anticline, a deposit of stratified rock beneath the surface of the Earth that folds gently downward, a little like the top of a tent or an inverted bowl. Anticlines exist because Earth's surface moves. These very slow and powerful motions sometimes cause layers of rock to bend. Sometimes the layers bend sharply, and sometimes they bend gently. When a

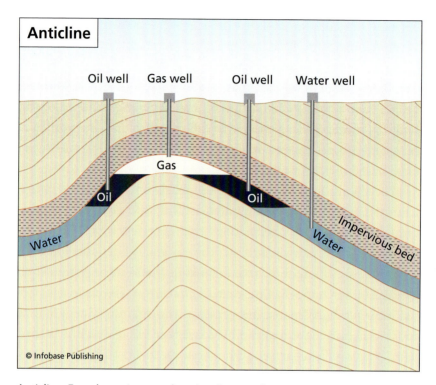

Anticline. For a long time, exploration for petroleum was almost synonymous with the search for anticlines.

layer of impermeable rock bends into an inverted bowl shape, it is called an anticline. Beneath the anticline are the reservoir rocks.

Driven by pressure differences, the slowly migrating gases and liquids may enter a nearby anticline (if an anticline is located nearby), and it is within the anticline that they become trapped. The water-petroleum mixture rises within the reservoir rock but is prevented from escaping by the impermeable rock around the reservoir rock. The accumulation process is slow but the length of time over which it occurs can be very long.

Because source rock, reservoir rock, and a trap (often an anticline) must all occur in proximity to one another, it is remarkable that any petroleum deposits form. But when the conditions are right, a great deal of natural gas and oil can accumulate within a large anticline. The conditions under which natural gas exists within an anticline depend on several factors. Sometimes the gas is dissolved in the oil—when oil is found with the gas—and in this situation, the gas and oil remain together until one of two conditions are met: Either the oil-gas mixture is drawn up through the well into a region of lower pressure, at which point the gas bubbles out of the oil, or the pressure in the reservoir drops—a pressure drop will occur as oil is removed from the reservoir, for example—at which point the gas and oil will separate while still within the reservoir rock. Alternatively, it is sometimes the case that the pressures and temperatures and the ratio of gas to oil within the reservoir are such that some or all of the gas, which is, after all, less dense than the oil, separates from the oil prior to the time that the drill pierces the anticline. Under these circumstances, the gas, which is more buoyant than oil, will migrate to the top of the anticline forming a so-called *gas cap*. Also, the oil will separate and rise above the water, which is denser than oil, with the result that the anticline will contain a three-layer deposit of gas, oil, and water, all under very high pressure.

Enormous quantities of natural gas may issue from a well once the drill has found its mark. By way of example, in 1947 the West Edmund Hunton Lime Pool field in Oklahoma was producing 300 million cubic feet (8.4 million cubic meters) of gas each day. In the rush to obtain the oil, however, all of this gas was simply vented or burned off as it came out of the well, a tremendous waste of natural gas that was more typical of the time than not.

The large oil- and gas-rich anticlines are also economically very attractive because they are often the easiest to exploit. They hold the promise of huge financial returns for the developers. Oil and gas companies have spent many years scouring the globe for these deposits, and they have probably drilled most of the really large anticlines—at least most of the easy-to-access ones. In the United States this is certainly true. If any large deposits of this type remain, they are probably located on federally protected lands where drilling is not permitted. Other geological formations have also been associated with deposits of gas. Salt domes—enormous masses of salt, sometimes miles across—may act to trap migrating gas and oil. Similarly, geological faults, places where the Earth's crust has fractured, are sometimes places where a mass of impermeable rock has slid across a region of more permeable reservoir rock, forming an effective trap. Within the United States, many of these formations have also already been identified and tested for oil and gas. Where else, then, are the future supplies of natural gas?

Increasingly, domestic suppliers seek to meet the ever-increasing demand for natural gas by searching areas that were previously too difficult to access. This explains why companies have found it necessary to develop the very expensive technology necessary to float drilling rigs a mile or so above the seafloor in the Gulf of Mexico in order to drill deep into the seabed, and it explains why these companies have also found it necessary to drill in ever more geographically remote locations and ever deeper beneath Earth's surface. The increasing scarcity of easy-to-reach deposits also explains the growing interest in so-called unconventional sources of natural gas, deposits of gas located in formations where the physics rather than the location makes the gas difficult to access.

An example of an unconventional source of natural gas is the *tight gas* reservoir. In tight gas reservoirs, the sands or rocks that contain the gas are insufficiently porous to allow gas to seep easily from one location to another—that is, the rocks may contain large amounts of gas, but there are too few channels in the rock through which the gas can flow from its original location to the well, even under the tremendous pressures that exist deep underground. It is estimated that more than one-fifth of all the recoverable natural gas in the United States is now located in tight gas formations, and so finding efficient and economical methods of extracting this gas is extremely important.

Until recently, tight gas deposits were not economical to develop. Until about 1996, for example, those operators drilling in the Bossier sands of East Texas would simply drill right through a tight gas formation in the hope of finding commercially valuable quantities of gas beneath it. The Bossier sands deposits, which are located between 12,000 and 15,000 feet (3,700–4,600 m) below the Earth's surface, were considered a drilling hazard rather than an economically valuable resource. The technology required to extract the gas from these sands was initially considered too expensive to justify the effort, but as the price of gas has risen that has changed.

One technique that has already been used to successfully extract gas from the Bossier Sands is called *hydraulic fracturing*. The technique is not new, but it has been significantly improved during experiments on the Bossier field. The idea is simple to explain, but technically difficult to accomplish: Hydraulic fracturing consists of injecting water or some other fluid into the gas-bearing rock at very high pressure in order to create fractures in the rock along which the gas can flow. Mixed with the water is particulate matter, often sand, to prop open the fractures. When the pressure is released the particulate matter becomes wedged within the fractures and prevents the fractures from closing entirely. (If the fractures were to reseal, the gas would once again be trapped inside the tight gas formation.)

Measurements at wells sunk in the Bossier Sands formation indicate that the improved hydraulic fracturing technique currently in use opened fractures with a typical length of approximately 400–500 feet (120–150 m). The particulate that was suspended with the fluid, however, seldom traveled more than about 250 feet (76 m) along the fracture before becoming lodged. Finding the right technique for extending the range of the particulates—the technical term for which is *proppants*—further into the fractures is a subject of intense experimentation. Hundreds of millions of dollars have been spent on finding economical ways to extract gas from the Bossier sands in a way that maximizes the amount of gas extracted.

Still another technique for extracting gas from tight gas reservoirs makes use of the fact that some vertical fractures already exist within any formation. This method involves innovations in drilling technology. In a tight gas field vertical fractures tend to be isolated. Until they are connected they remain of little value to those attempting to develop the gas field because although they may connect the lower part of the field with the upper part, they provide no path by which the gas can be extracted since they begin and end within the field. The goal of the drillers, then, is to connect as many of these vertical fractures as possible. To this end, wells are drilled either at a slant through the gas-bearing rock or sometimes down to the gas reservoir and then horizontally along the gas field. As the drilling proceeds more and more vertical fractures are connected, thereby increasing the gas flow rate.

There are a number of other techniques and ideas under development. Some involve drilling to great depths; others involve recovering methane from coal deposits, where until recently miners had perceived methane solely as a safety hazard rather than a resource. (This is called coal-bed methane, and it is already making a modest contribution to the gas supply of the United States.) When successful, each innovation opens up still more gas reserves for exploitation, an extremely important consideration because so much electricity production and home and commercial heating depends upon a reliable and reasonably priced supply of natural gas.

The conventional and unconventional sources described so far in this chapter are the source of most of the natural gas—that is, most of the methane—produced around the world. But most of Earth's methane is found elsewhere, in an ice-like material called methane hydrate. Methane hydrates are composed of methane molecules surrounded by water molecules. Together, at the right temperature and pressure, the two materials form a solid that looks much like ordinary ice. The most striking difference between methane hydrates and water ice is that when ignited, methane hydrates will burn. As with ordinary ice, methane hydrates cannot persist in warmer climes. At higher temperatures methane hydrates "dissociate," meaning that the methane molecules separate from the water molecules, producing methane gas. But methane hydrates can exist at temperatures somewhat above freezing if the pressure is high enough. The temperatures and pressures at which methane hydrates exist define a region of Earth called the *hydrate stability zone*: It includes parts of the Arctic tundra and long, wide bands beneath the oceans along the margins of the continents at depths exceeding 1,000 feet (300 m). Although there are substantial accumulations of hydrate beneath the tundra, by far the largest deposits are on and beneath the sea floor. Some hydrate is found as isolated nodules on the floor of the ocean, but most of it is found beneath the seabed in the pores of rocks and mud in vast deposits as much as 1,000 feet (300 m) thick. Because the methane molecules in hydrates are packed more closely than they would be in a gas, methane hydrate deposits contain large amounts of methane per unit volume. When a unit volume of methane hydrate dissociates, it can produce as much as 160 units of pure methane gas at atmospheric temperatures and pressures.

There is a great deal of uncertainty with respect to the amount of methane hydrate that exists, but all estimates point to staggering quantities: In Alaska and along the East, West, and Gulf coasts of the United States, for example, the best available data indicates reserves in the neighborhood of 200,000 trillion cubic feet (5,600 trillion cubic meters) of methane, a number that is written as a 2 followed by 17 zeroes. That is enough natural gas to satisfy current levels of U.S. demand for centuries (assuming a way is found to produce it). Nor is the United States alone. Japan, Canada, India, and many other countries with long coastlines have discovered enormous deposits of hydrates within their territorial waters, and the United States, Canada, and Russia also have large reserves of methane hydrates locked in the tundra.

There are no hydrate wells currently in production. Unlike oil and gas, hydrates do not flow. Consequently, one cannot simply sink a well into a deposit of methane hydrates and pump out the contents. Nor can methane hydrates be mined as coal is mined since they will dissociate at ordinary temperatures and pressures. Further complicating recovery efforts, most deposits are located far beneath the sea.

In theory, it is possible to exploit deposits of methane hydrates by depressurizing the deposits or by heating them. In each case, the methane hydrates would dissociate and, in theory, allow the well operator to capture the methane for use as a fuel. It is also, in theory, possible to inject a chemical to dissolve a deposit of methane hydrate and recover the resulting gas. The chemical would work in a way that is similar to the deicing mixtures sometimes sprayed on the wings of airplanes. None of these techniques are ready for commercial application, and so it remains to be seen how much hydrate can be produced by any of these methods or any combination of them. But even if only one percent of all the methane hydrates within the United States were commercially recoverable, it would still be enough to double the nation's supply of natural gas.

Methane hydrates are also of interest for another reason entirely. Methane is a powerful *greenhouse gas,* a gas that, when added to the atmosphere, can cause the Earth's temperature to increase. Methane remains in the atmosphere for a shorter period of time than carbon dioxide, but its effect is roughly 20 times greater than that of CO_2—in other words, methane molecules in the atmosphere at a *density* of x parts per million, where x represents an arbitrary level of methane concentration, have the same effect on the atmosphere's ability to retain heat as CO_2 molecules at a density of about $20x$ parts per million. If a significant fraction of the methane present in all of the deposits of methane hydrates were to be released into the atmosphere, the density of methane in the atmosphere would increase substantially, possibly with disastrous effects. This is important because as the global climate warms, the methane hydrates that are present in the Arctic may begin to dissociate, releasing large volumes of methane into the atmosphere. But whether this would actually happen is a subject of current research.

Interestingly, NASA has discovered evidence that 55 million years ago, because of movements of the Earth's crust, large deposits of methane hydrates located on the seafloor dissociated, releasing enormous quantities of methane gas into the atmosphere. Some scientists claim that the result was an increase of 13°F (7°C) in Earth's average temperature. The relationship between methane hydrate dissociation and climate change is, however, controversial and it, too, remains the subject of research and debate.

SYNTHETIC GAS

Fossil fuels are sometimes classified according to their carbon-to-hydrogen ratios. The carbon-to-hydrogen ratio is formed by comparing the mass of carbon present in a sample of the fuel in question to the mass of hydrogen present. As fossil fuel technology has evolved, humanity has found ways to use fuels with lower carbon-to-hydrogen ratios. Coal, for example, which was the first fossil fuel to be used, has been exploited for centuries. Bituminous coal, a commonly used type of coal, has a carbon-to-hydrogen ratio that ranges from 14:1 to 18:1, depending upon the sample. Kerosene, a fuel that dates to the latter half of the 19th century, has a carbon-to-hydrogen ratio of 6.3:1, and methane, which came into use even later than kerosene, has a carbon-to-hydrogen ratio of only 3:1. The comparison between coal and natural gas is especially important because most of the electricity in the United States is produced from only three types of power plants: coal-fired, natural gas–fired, and nuclear plants.

One advantage of using fuels with lower carbon-to-hydrogen ratios is that they tend to burn more cleanly. In particular, consumption of fossil fuels with lower carbon-to-hydrogen ratios cre-

The Great Plains Synfuels Plant converts coal to methane and produces a number of commercially valuable coproducts in the process. *(Basin Electric Power Cooperative)*

ates less CO_2 per unit of heat produced. Among the fossil fuels, natural gas has the lowest carbon-to-hydrogen ratio, and it is also the cleanest burning in the sense that it produces the fewest harmful emissions per unit of thermal energy produced. The reduction in emissions levels that a power producer achieves by using natural gas rather than, say, coal, has substantial economic as well as environmental advantages. It is *because* natural gas–fired power plants are so clean that they are, at least in the United States, easier to site and cheaper to build than coal plants that produce equal amounts of power.

An important disadvantage, relative to coal, of using natural gas is that, on average, natural gas prices are high and are expected to remain high. The price of gas from newly developed gas fields reflects the greater costs involved in producing it and the strong demand for gas. Nor have domestic supplies kept pace with increasing demand. Imports of liquefied natural gas, which is brought into the United States in specially designed tankers, are rising, and liquefied natural gas also tends to be expensive. As a consequence, power producers considering building new fossil fuel plants, which have lifetimes measured in decades, are faced with choosing between natural gas, a cleaner-burning but expensive fuel that may be even more expensive in the future, or coal, which is both dirtier and much cheaper than natural gas.

For the consumer a further disadvantage to using natural gas is that supplies of natural gas are subject to disruption. In 2005, for example, Hurricane Katrina disrupted supplies of gas from the Gulf of Mexico, causing brief shortages and price spikes throughout the nation. By contrast, because coal is so abundant in the United States, its price is not as volatile as that of natural gas nor is the supply subject to disruption. (Very little oil is burned to produce electricity anymore because the price is too high and unpredictable.)

When choosing between building a coal-fired or natural gas–fired power plant, there is, at least in theory, a third alternative. It is possible to manufacture natural gas from coal. This has already been accomplished. This synthetic natural gas is almost pure methane. Since the 1980s the federal government has subsidized an enormous synthetic gas plant called the Great Plains Synfuels Plant, located in Beulah, North Dakota. The plant is a giant experiment, the goal of which is to produce inexpensive synthetic gas from an almost inexhaustible supply of inexpensive coal. Conceptually, the plant is, in some ways, similar to the coal gas plants that were found throughout the nation during the 19th and early 20th centuries, but in contrast to coal gas, which was dirtier and had a *lower heating value* than natural gas, the fuel produced by the Great Plains plant is methane, which is chemically identical with the principal constituent of conventional natural gas.

The Great Plains plant, which began operation in 1984 and which, even today, continues to consume 6 million tons of coal per year, was built near an enormous deposit of lignite, also known as brown coal. Originally designed to produce 125 million cubic feet (3.5 million cubic meters) of pipeline-quality natural gas each day, it was upgraded in 2004 and now successfully manufactures in the neighborhood of 170 million cubic feet (4.8 million cubic meters) of natural gas per day.

But the plant does more than produce natural gas. Chemically, coal is an extremely rich resource, and the process used to manufacture methane also produces a number of other valuable coproducts. The Beulah plant has a yearly output of 400,000 short tons (360,000 *metric tons*) of anhydrous ammonia, a commonly used fertilizer, 7 million gallons (26 million liters) of naphtha, which is used in gasoline and in the production of solvents, and a number of other commercially valuable products. The process also produces 200 million cubic feet (5.6 million cubic meters) of CO_2 each day. The huge volumes of carbon dioxide produced by the plant reflect the fact that it is turning lignite, a material with a high carbon-to-hydrogen ratio, into a fuel with a much lower carbon-to-hydrogen ratio. The excess carbon shows up as CO_2, which, initially, was just vented into the atmosphere. Today, however, a 205-mile (330-km) CO_2 pipeline stretches from the Great Plains Synfuel Plant to oil fields in Saskatchewan, Canada, where it is injected into the oil fields to increase field pressure and improve oil recovery rates. The North Dakota plant is paid for this CO_2. (It is sometimes asserted that use of CO_2 in oil fields "sequesters" the carbon dioxide, but the purpose of injecting the CO_2 is to increase the rate at which oil is recovered by increasing the oil field pressure. There has not been enough study to confidently state that CO_2 injected into an oil field in this way will not eventually make its way to the surface.) Despite the large volumes of methane and commercially valuable coproducts produced by the plant, the Great Plains plant has historically been unprofitable to operate. It remains a work in progress, dependent on government subsidies, a project from which engineers continue to learn, and a reminder of how difficult it is to solve many energy problems.

The Great Plains Synfuel Plant was the first modern attempt to make a synthetic natural gas. A number of synthetic gas plants have since been built, but the concept of what a synthetic gas plant should produce has changed. For example, plants in Mulberry, Florida (the Polk Power Station), and Terre Haute, Indiana (the Wabash River Clean Coal Power Plant), have been built to manufacture a combustible gas from coal, but rather than methane, the synthetic gas (syngas) produced at these plants is largely a combination of hydrogen gas and carbon monoxide, with smaller amounts of water vapor and carbon dioxide. At these newer plants, there is no need to produce a synthetic fuel that is interchangeable with natural gas, because in contrast to the Great Plains plant, all of the gas manufactured at the Florida and Indiana plants is consumed on-site in specially designed facilities for the production of electricity—that is, there is an electric generating station at each site and the gas manufactured there is only used to supply the fuel needs of the generating station. In contrast to coal plants, which produce electricity by converting liquid water into steam, which is then used to drive a steam *turbine,* the newer coal gasification plants use an integrated gasification combined cycle (IGCC), a technology that converts more of the thermal energy into electricity.

The "CC" in IGCC technology refers to a two-step, "combined cycle" process. First, the manufactured gases are burned, causing them to expand and push against the blades of a specially designed

(continues on page 40)

Carbon Dioxide Sequestration

Burning natural gas (and other fossil fuels) produces carbon dioxide (CO_2) as a by-product. This cannot be avoided. Heating, transportation, and electric power production release billions of tons of CO_2 each year. At present, virtually all the resulting CO_2 is vented directly to the atmosphere. "The current atmospheric carbon dioxide concentration is approximately 380 ppm [parts per million] volume and rising at a rate of approximately two ppm volume annually," according to Robert C. Burrus, a geologist with the U.S. Geological Survey, in 2007 testimony before the Senate Subcommittee on Science, Technology, and Innovation. As CO_2 levels rise, the heat-retention properties of the atmosphere change, and the atmosphere retains more of the Sun's energy. The result is that, on average, Earth's temperature is somewhat higher than it would otherwise be. The change in temperature is not large, but its effects (some of which are described in chapter 4) are. What can be done?

Conceptually, there are several approaches to controlling global temperature change. Some of the more talked-about approaches involve the following:

1. switching fuels,
2. becoming more efficient (and so using less energy to achieve the same result),
3. removing the CO_2 from the atmosphere after it has been emitted,
4. changing the reflective properties of the atmosphere so that the Earth receives less sunlight, and
5. removing the CO_2 produced by combustion before it reaches the atmosphere.

This last approach leads to the idea of *carbon sequestration*.

Carbon dioxide is a gas at ordinary temperatures and pressures. Commercially, it has little value because the demand for CO_2 is dwarfed by the supply. As a consequence, it must be emitted into the atmosphere or hidden; the technical term for "hidden" is *sequestered*. The idea behind sequestration technology is to place the gas where it will remain "permanently" trapped—in a coal mine, within the ocean's depths, or injected deep beneath the Earth's surface.

For technical reasons, no one is presently contemplating the possibility of sequestering the CO_2 produced in the transportation sector or in the residential heating market. This leaves the power generation sector as the one place where sequestration technology, if it is implemented at all, will be introduced. But to have an effect on climate change, sequestration technology must be implemented on a large enough scale to hide billions of tons of CO_2 indefinitely.

Sequestration technology is expensive. Some of the biggest CO_2 emitters, China and India, have shown little interest in implementing sequestration technology, preferring instead to apply scarce dollars to projects with more immediate effects on the welfare of their citizens. And although several developed nations, such as the United States and Germany, have conducted a good deal of research on the problems associated with sequestration, they have yet to implement it on a large scale. The reason is that sequestration is hard to accomplish, and it is local in the sense that good sequestration sites depend on the local geology. Absent a huge pipeline infrastructure for transporting the CO_2 produced at power plants to distant sequestration sites, which would be one more huge cost in an already expensive system, the CO_2 produced by each power plant will be sequestered nearby. Each site must, therefore, be studied individually to determine the best way of sequestering the CO_2 produced *at that site*. Even under the most optimistic assumptions, therefore, sequestration technology will probably not be in widespread use for decades. What is to be done with the CO_2 that results from fossil fuel consumption? The answer is not yet clear. The truth is that at present humanity cannot live without combustion technology, but is finding it increasingly hard to live with it.

(continued from page 37)

gas turbine, a device for converting energy in the rapidly expanding combustion gases into rotary motion. After the hot gases pass
over the blades of the gas turbine they are directed towards a *heat
exchanger,* which transfers some of the thermal energy in the hot
gases to water, which turns the water into steam. The steam is used
to turn a second turbine, a steam turbine, which converts the linear
motion of the expanding steam into rotary motion. The turbines
turn *generators.* The generators convert rotary motion into electrical energy. The ultimate purpose of all the hardware in the plant is
to turn the generators. The value of IGCC technology is that a combined cycle plant can convert upwards of 50 percent of the thermal
energy produced at the plant into electrical energy. A more conventional coal plant, which uses only a steam turbine, will convert
only about one-third of the thermal energy produced at the plant
into electricity. Higher efficiencies mean that less fuel is required,
and this should, in theory, reduce the environmental effects of plant
operation by reducing the fuel consumption required to produce a
given amount of power.

In addition to a higher degree of efficiency, gasification enables
the plant operators to further reduce emissions by manufacturing
a cleaner-burning fuel than the pulverized coal in common use
today. By way of example, the syngas manufacturing process enables the plant operators to capture much of the sulfur present in
the coal as they are manufacturing the syngas. Because the gaseous
fuel is nearly sulfur-free, so are the emissions produced by burning
it. (Both the Florida and Indiana plants still vent their CO_2 into the
atmosphere.)

These two syngas plants represent significant conceptual and
technical progress in terms of finding ways to burn coal cleanly.
They offer the possibility of displacing at least some natural gas
plants with plants that use coal as a primary energy source but
produce emissions that are more like those produced at natural

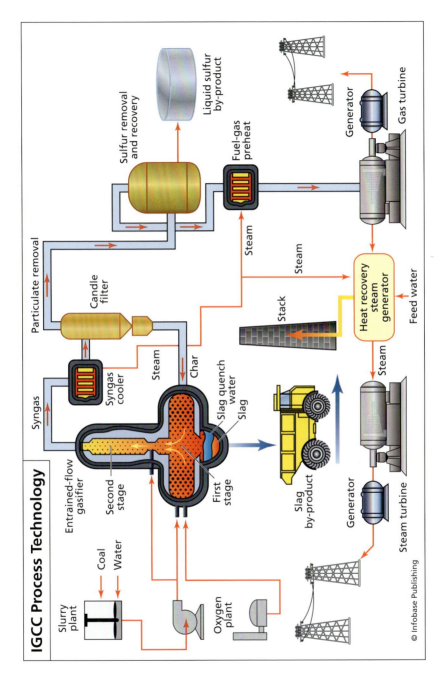

IGCC Process Technology

IGCC process technology offers the possibility of coal-based, gas-fired power plants with greatly reduced emissions. *(U.S. Department of Energy)*

gas–fired plants, except for substantially higher emissions of CO_2. In addition to their potential environmental advantages, there are economic advantages to replacing natural gas with clean-burning coal. Since the 1990s, United States power producers have met increased demand for electricity by building a great many natural gas–fired plants. The newly constructed natural gas–fired plants have increased the demand for natural gas and contributed to a rapid increase in natural gas prices. A fleet of plants like those at Mulberry and Terre Haute would, of course, have no such effect. Despite their promise, however, investors have so far shown little interest in building these more advanced power plants, which tend to be more expensive to build and less reliable to operate than their more conventional counterparts. Additional research is needed before this technology becomes the technology of choice.

FutureGen, a project supervised by the U.S. Department of Energy, was the latest attempt to demonstrate the value of coal gasification and IGCC technology. A public-private partnership involving participants from around the world, FutureGen was an attempt to convert coal into electricity and pure hydrogen gas, an energy-rich pollution-free fuel. The carbon dioxide generated at the plant would have been captured and sequestered. After years of publicity by the Department of Energy, the project was abruptly cancelled in January 2008. The government cited cost overruns. The projected cost of the project had increased from about $1 billion to $1.8 billion. The cancellation of the project was controversial because of the project's potential importance. Many of the new technologies needed to greatly reduce the environmental impact of the next generation of coal-dependent power plants were to have been demonstrated by the FutureGen project. Commercial-scale tests of these technologies have now been delayed for at least another several years.

In the United States the easy-to-access natural gas is gone. Roughly 70 percent of all natural gas currently produced comes from wells that exceed 5,000 feet (1,500 m) in depth. Increasingly, natural gas will come from very deep deposits of gas—in excess of 15,000 feet (4,600 m)—and tight gas formations. Certainly coal-bed methane will make a somewhat larger contribution to the nation's gas supply. Additional natural gas can be manufactured as is done in North Dakota, and various types of syngas can, in theory, be used in place of natural gas at power plants. (All of this effort will prove unnecessary if a way is found to produce commercial amounts of methane from deposits of methane hydrates.) Technology can make an enormous difference with respect to the natural gas industry, not just in terms of how natural gas is used but also with respect to the amount of natural gas and natural gas–like fuels available for use.

Transporting and Storing Natural Gas

Historically, the natural gas business has evolved in ways very different from those of the coal and oil businesses, in part because natural gas, as a gaseous fuel, requires a transportation infrastructure different from that of coal or oil. To appreciate the natural gas business, therefore, it is important to appreciate some of the unique aspects of the natural gas transportation infrastructure. Describing that infrastructure is one of the goals of this chapter.

Demand for natural gas fluctuates from day to day depending, for example, on the weather. Spikes in demand can tax the ability of the natural gas transportation infrastructure to provide natural gas when and where it is needed. To reduce the possibility of bottlenecks in the system, suppliers have developed interesting and creative ways to store huge amounts of natural gas safely and at a modest cost. Describing the natural gas storage system and its effects on supplies and prices is the other goal of this chapter.

GAS PIPELINES

The main method by which natural gas is transported from one place to another is by pipeline. Because gas fields are often located far from those willing to pay for it, vast pipeline networks have been constructed connecting, for example, gas fields in Russia with consumers in Germany and gas fields deep beneath the Gulf of Mexico with consumers in New England. Such a system is extremely expensive to build. The worldwide system of natural gas pipelines has been described as a $2 trillion network, and each year many trillions of cubic feet of natural gas flow through the network. Pipelines are indispensable. They are the only method available for transporting such large volumes of gas.

Although many nations have built extensive pipeline networks, the United States has by far the largest, with approximately 329,000

Welding pipeline. The early development of the natural gas business was inhibited by inadequate pipeline technology. *(Enbridge Inc.)*

miles (526,000 km) of natural gas pipeline. This total includes the 212,000 miles (339,000 km) of interstate pipeline network, the intrastate pipeline networks—which total approximately 76,000 miles (122,000 km) of pipeline—and 41,000 miles (67,000 km) of field gathering lines. (This total does not include hundreds of thousands of miles of regional pipelines, nor does it include the pipelines used in local distribution systems. These additional systems would raise the total to more than 1 million miles of pipeline.) By comparison, Russia is second with approximately 95,000 miles (150,000 km) in its pipeline network. In this section, unless specifically noted, the details of the pipeline technology described are those of the United States system. Other systems are, however, similar in many ways.

The actual pipes that comprise a natural gas pipeline generally fall into one of two categories, main lines or lateral lines. Main lines are used for interstate distribution of natural gas and, as a general rule, have diameters between 16 and 48 inches (41–122 cm). They are connected to the hundreds of thousands of natural gas wells in the United States by so-called lateral lines, which have diameters in the 6–16 inch (15–41 cm) range. Lateral lines also distribute gas from main lines to local distribution systems.

Constructing the sections of pipe that together constitute a pipeline is an enormous and very technical undertaking. As mentioned in chapter 1, the absence of the technology required to manufacture large diameter pipes was one of the impediments to the creation of an early natural gas industry. Today, large diameter pipes are constructed from rectangular sheets of high-strength carbon steel. These are rolled into cylinders and then welded along the seam. They are protected by special coatings, and examined with an array of technologies to ensure that they will function as intended.

Because pipelines are expensive, pipeline companies expend a good deal of effort learning how to use them as efficiently as possible. To this end, companies seek to transport the maximum amount of gas per pipeline. To accomplish this, they put the gas under pres-

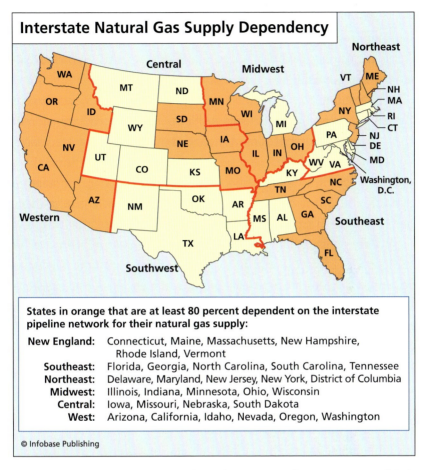

Interstate Natural Gas Supply Dependency

States in orange that are at least 80 percent dependent on the interstate pipeline network for their natural gas supply:

New England: Connecticut, Maine, Massachusetts, New Hampshire, Rhode Island, Vermont
Southeast: Florida, Georgia, North Carolina, South Carolina, Tennessee
Northeast: Delaware, Maryland, New Jersey, New York, District of Columbia
Midwest: Illinois, Indiana, Minnesota, Ohio, Wisconsin
Central: Iowa, Missouri, Nebraska, South Dakota
West: Arizona, California, Idaho, Nevada, Oregon, Washington

© Infobase Publishing

On the map, the orange states are those that receive at least 80 percent of their natural gas through the nation's interstate pipeline system, one more indication of how geographically extensive the system is and how much gas flows through it. *(Energy Information Administration)*

sure. This reduces the volume of the gas and enables the pipeline company to transport more gas (and so more energy) per pipeline per unit time. Although the amount of pressure varies from pipeline to pipeline and is a function of demand and pipeline design, typical pipeline pressure is about 70 atmospheres—that is, the gas inside the pipeline is maintained at about 70 times the atmospheric pressure at sea level. Older pipelines may operate at a somewhat lower pressure, and the proposed Alaska Natural Gas Transportation System will, if it is completed, have a maximum allowable operating pressure of at least 170 atmospheres.

The volume of gas handled by these pipelines is extraordinary. By way of example, the Cheyenne Plains Pipeline, an extension of the Colorado Interstate Gas system, began operation in December 2004 and has a capacity of 560 million cubic feet (16 million cubic meters) of gas per day, and the Alaska Natural Gas Transportation System will, if it is built, have a capacity of at least 4 billion cubic feet (112 million cubic meters) of gas per day.

To keep the gas flowing at the designated pressure, compressor stations are situated every 40 to 100 miles (60–160 km). The pressure must be restored, or at least periodically checked, because some of the so-called natural gas liquids, heavier hydrocarbons such as butane, may enter a pipeline as gases mixed with the methane. Under the right conditions, some of these heavier hydrocarbons will condense within the pipeline. Condensation of a gas results in a drop in pressure. Water vapor, present as a contaminant, may also condense. These liquids are separated from the gas stream at the compressor station, and the system pressure restored.

Interspersed with the compressor stations are valves, which are located every five to 15 miles (8–24 km) along a pipeline. These enable operators to close down sections of the pipeline for inspection or repair. In many ways, pipelines are operated in a manner that is analogous to that of an electric grid. The compressor stations are analogous to transformers, and valves enable pipeline operators to shut down sections of the network just as technicians are able to shut down sections of the electrical grid for repair or maintenance. Even the regulations that govern operation of interstate pipeline networks are somewhat similar to those that govern the operation of high-voltage transmission networks. (See chapter 6 for information on the way that natural gas is regulated in the United States today.)

(continues on page 50)

The Energy Content of Natural Gas

"Natural gas pipeline companies prefer to operate their systems as close to full capacity as possible" according to the Energy Information Administration. This rate is expressed in terms of the volume of gas transported per unit time. It is, however, just as important to know how much energy is transported per unit time. What, in other words, is the energy content of natural gas?

The energy content of natural gas is often expressed as the thermal energy released by burning one *standard cubic foot* (scf) of natural gas. (A standard cubic foot of natural gas is the amount of gas required to fill one cubic foot [0.028 m³] of space at atmospheric pressure at 60°F [15.6°C].) This energy value is determined in the laboratory under carefully controlled conditions. There are two closely related methods of expressing the energy content of a fuel: the lower heating value (LHV) and the

(continues)

This automobile runs on compressed natural gas. With the right hardware and a natural gas connection it can be refueled at home. *(Honda Media Newsroom)*

(continued)

higher heating value (HHV). To understand the difference, keep in mind that burning natural gas produces carbon dioxide (CO_2) and water (H_2O) in large quantities and very small amounts of other materials. In fact, a combustion reaction that produces only CO_2 and H_2O is exactly what is meant when one says that a fuel burns cleanly.

The difference between the LHV and HHV is determined by the phase of the water produced by the combustion reaction at the time that the measurement is made. Water will be present as either steam or as a liquid. The phase of the water is important because a good deal of thermal energy is released when steam condenses to form liquid water. The energy released during the phase change from gas to liquid is called the latent heat of condensation. It does not change the temperature of the water, but it does represent a substantial flow of heat into the surrounding environment. If, when the laboratory measurement is made, the water produced by the reaction is in the form of a vapor, one obtains the LHV. If, on the other hand, the water is in the form of a liquid, one obtains the HHV. In describing the heating value of a fossil fuel, the lower heating value is often used because an engine does not make use of the latent heat of condensation.

The LHV of natural gas is commonly given as about 930 Btu per cubic foot (34.6 MJ/m³), and the HHV for natural gas is approximately 1,027 Btu per cubic foot (38.3 MJ/m³). For purposes of comparison, the LHV of gasoline is commonly given as 115,000 Btu per gallon (32 MJ/liter), and the

(continued from page 48)

LIQUEFIED NATURAL GAS

Even the most sophisticated pipeline technology has its limitations. Pipelines cannot span oceans, and it is sometimes not economical to use pipelines to connect markets and producers, even when, in

HHV of gasoline is approximately 125,000 Btu per gallon (34.8 MJ/liter). (All of these values vary somewhat from sample to sample.) It is clearly difficult to compare the energy content of these two fuels when the LHV and HHV are given in this way. For example, how much natural gas must be consumed to provide the same amount of thermal energy as one would obtain by burning a gallon of gasoline? To answer this question one must use the same units for both sets of heating values.

If one changes the heating values for both natural gas and gasoline to energy released per unit mass, one finds that the heating values of these fuels are similar. The main difference between them, therefore, is that one is gaseous, and one is a liquid. This is important because at room temperature and at a pressure of one atmosphere, one unit mass of natural gas occupies a volume that is more than 1,000 times as large as the same mass of gasoline. As previously mentioned, the volume of gas that is transported can be reduced somewhat by transporting it under pressure, but this does not change the fact that there does not exist much energy in a cubic foot of natural gas even under pressure. Therefore, meeting the demand for energy with natural gas requires the transmission of enormous volumes of gas. The United States, for example, regularly consumes in excess of 22 trillion cubic feet (620 billion cubic meters) of natural gas each year, which is approximately 150 cubic miles (620 cubic kilometers) of natural gas at standard temperature and pressure.

theory, such a connection is possible. Sometimes the distance involved is too great; sometimes the topography is too difficult, and sometimes there are concerns about the physical security of the pipeline. For many years, therefore, there were no markets for natural gas produced in certain inaccessible locations because there was no way to get the gas to market. If the gas found in these regions was

LNG tanker *(Federal Energy Regulatory Commission)*

not also found in conjunction with oil, it was ignored. If, however, it was found in conjunction with oil, it was often flared—vented into the atmosphere and burned in the open air to prevent it from becoming a safety hazard. Enormous quantities of gas were wasted in this manner. But the situation has now changed. Natural gas pipeline networks are now supplemented by liquefied natural gas (LNG) tankers, ships specifically designed to carry natural gas.

The LNG process works as follows: Natural gas is produced at a location that has no pipeline connection to a market. Instead, the gas is transported via pipeline to a seaport with a liquefaction plant, a highly specialized facility—they cost upward of $1 billion apiece—designed to remove impurities from the gas and then to cool the gas to a temperature of –260°F (–162°C), at which point the methane condenses into a liquid. The LNG is stored at atmospheric pressure in huge, heavily insulated tanks until it is loaded onto LNG tankers, which are specially built ships that transport the LNG to ports that are equipped with the necessary facilities to offload this very special, very cold cargo. (There are currently only four such locations in the United States.) Upon arrival, the content of the LNG tankers is loaded as a liquid into heavily insulated tanks, where it is held until the operators are ready to run it through a facility that raises the temperature of the liquid, thereby converting it back into a gas, which is then injected into a pipeline and transported to market.

LNG tankers typically have a capacity of about 5 million cubic feet (140,000 m³), which means that each tanker carries the equivalent of about 2.9 billion standard cubic feet (81 million m³) of natural gas. As with the liquefaction plants, LNG tankers are also very expensive. In fact, there is nothing about this process that is cheap, which raises the question of why the LNG process is so attractive to so many investors. The advantage of using a liquefaction process is that it reduces the volume occupied by natural gas at room temperature and atmospheric pressure by a factor of approximately 600. A single LNG tanker can, therefore, contain sufficient liquefied natural gas to supply roughly 5 percent of the average one-day demand in the U.S. market, a remarkable engineering achievement.

The LNG business is relatively new. It began in 1959 when the first LNG tanker ever built carried its cargo from Louisiana to Great Britain. Today, there is a modest but growing market for LNG in the United States and Europe. There is a much larger market in Asia. Japan, for example, obtains almost its entire gas supply in the form of LNG. Taiwan and South Korea are also important importers of LNG. Indonesia, which supplies these markets, has long been the largest exporter of LNG, producing roughly one-fifth of the world's total LNG supply.

LNG is just a form of natural gas, chemically identical with the gaseous form of the fuel. Nevertheless, siting LNG plants in the United States has sometimes proved very controversial. The reason most often cited is safety. To be sure, LNG tankers and LNG storage facilities contain tremendous amounts of energy, which, if it were released in an uncontrolled manner, could devastate the area of the accident. In evaluating the *actual* risk, however, it is worth noting that the safety record of the LNG industry is quite good. In more than 40 years, LNG tankers have transported more than 33,000 shipments of LNG without a single serious accident. LNG facilities have safety records that are also excellent. A great deal of ingenuity has been expended making all aspects of the LNG process as safe as possible. For example, the containment system on a modern LNG tanker is more than six feet (1.8 m) thick, and in the United States, while an LNG tanker is entering port and while it is docked, the Coast Guard does not allow any ships to approach it.

It is certain that the contribution of LNG to the global natural gas supply system will continue to grow. Russia, the world's biggest producer of natural gas, is aggressively expanding its LNG market. On the consumption side, the U.S. Energy Information Administration (EIA) predicts that, provided the cost of natural gas remains high enough, imports of LNG into the United States will rise steadily from 2005 through 2025, at which time LNG imports will have more than quadrupled and will constitute 7 percent of the nation's total gas supply.

NATURAL GAS STORAGE

Since the end of World War II, a sophisticated system for storing large quantities of natural gas has developed as suppliers have attempted to better balance supply and demand. The natural gas industry is a seasonal business. It is not as seasonal as it once was, because large numbers of new natural gas–fired power plants have greatly increased demand for gas during the warm summer months. Nevertheless, demand continues to fluctuate, going down during the summer and up in the cold winter months as a result of the heating market.

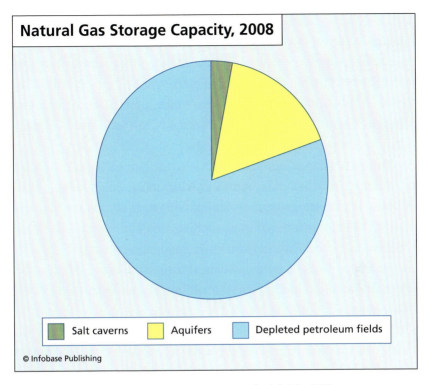

Natural Gas Storage Capacity, 2008

Salt caverns   Aquifers   Depleted petroleum fields

© Infobase Publishing

Most natural gas storage is in depleted gas and oil fields. *(EIA)*

To meet even the highest levels of wintertime demand, pipeline operators could expand the capacity of the pipeline network to transport sufficient natural gas from the fields where it is produced to the often-distant markets where it is needed. But such an approach would be very expensive and would result in an overbuilt network—overbuilt in the sense that the network would be underutilized practically all of the time. Pipeline operators would have had to pay for capacity that they rarely needed. In order to avoid this kind of inefficiency, pipeline operators developed an extensive system of storage facilities located near major markets in order to keep sufficient quantities of gas on hand to meet periods of high demand. The idea is simple: Fill up the storage facilities during periods of low demand, when the pipeline network has lots of spare capacity, and draw down supplies during periods of high demand, when the pipeline system is working at capacity. Because the storage facilities are located near the intended markets, this gas can be brought to market without using the main parts of the pipeline network. By carefully choosing where the gas storage facilities are located, pipeline companies are better able to efficiently balance supply with demand when demand is high.

The natural gas storage infrastructure in the United States has the capacity to hold about 4 trillion cubic feet of natural gas in about 415 underground storage facilities. It is evident that the capacity of at least some of these facilities must be very large. By far the most common way to store large quantities of natural gas is to transport gas to depleted gas fields and then to inject gas into the depleted field. Essentially, this is the natural gas business operated in reverse.

There are several reasons why depleted gas and oil fields, provided that they are located near major gas markets, are ideal places to store large quantities of natural gas. Petroleum engineers have already studied these fields in great detail. They know how much gas each field can hold, and they know the permeability of reservoir rock in each field. (The permeability of the rock determines how fast gas can move through the formation, which determines the rate at which gas can be injected or withdrawn from the field.) A depleted field can be accessed through the wells that had once been used to extract gas, and the storage facility's operators can connect their facility to the pipeline network by using the pipeline system that had been previously built to service the field when it was first productive. Finally, when a gas field is abandoned, there is always some gas left in it. It was abandoned when the reservoir pressure had dropped so far that the remaining gas did not flow to the well. Any gas that is injected back into the field raises the pressure of the field back above what is necessary to make the gas flow. Essentially, a depleted field is already "primed." When operators use other types of geological formations, they must begin by injecting gas—so-called base gas, gas that they will never get back—in order to establish a minimum field pressure. This practice represents an additional expense that can be avoided by using a depleted natural gas field.

There is one disadvantage to using a depleted field and that is that the rate at which gas can be injected or withdrawn from the field is not very fast. Even the most permeable reservoir is, in the end, mostly rock. Another type of reservoir, called a salt cavern, provides a storage facility that enables operators to add and withdraw gas quickly. Some salt deposits, called salt domes, are a mile (1.6 km) or more across and extend miles into the Earth. Although they are solid, it is relatively easy to form caverns within them. The process begins by drilling deep within the salt dome—anywhere from 1,000 to 6,000 feet (300–1800 m)—and then pumping water down the hole. As the water circulates it dissolves some of the salt, and when the water is removed so is the dissolved salt. What remains behind is a void or cavern within the formation. It is in this void that the gas is stored.

A salt cavern cannot hold nearly as much gas as a depleted field. Its value lies in the fact that operators can add or withdraw gas very rapidly. This characteristic enables operators of salt caverns to respond more quickly than operators of depleted fields. Salt caverns fill a niche not filled by depleted gas fields.

Gas can also be stored in aquifers, geological formations that are saturated with water, provided that an impermeable layer of rock overlays the aquifer to prevent the gas from escaping. Aquifer storage facilities have been developed in places where there are no depleted wells, but the cost of developing aquifer storage is high, and government regulations intended to protect groundwater have further complicated efforts to develop more such facilities.

The difference between depleted gas fields and salt caverns is sometimes summarized by saying that gas fields are for base load storage and the salt caverns are useful for peak storage. Base load storage is to accommodate seasonal fluctuations in demand. The amount of gas in a base load storage facility fluctuates slowly and in accordance with the seasons. The amount of gas in the facility is lowest at the end of winter and, because gas is injected into the facility throughout the warmer months, storage levels are highest just before winter begins. These kinds of fluctuations are well suited to the permeability of the reservoir. By contrast, the levels of gas in a salt cavern rise and fall more chaotically. There might be a drawdown to compensate for a break in a pipeline compressor station, and levels might spike upward if there is a drop in natural gas prices, and operators want to buy low so that they can sell high later.

Electricity Generation

The main uses of natural gas are for residential, commercial, and industrial heating, process heat, and electricity generation. Other applications exist in the petrochemical industry, which uses natural gas as a *feedstock* or raw material in various manufacturing processes, and in the fertilizer industry, which uses natural gas in the production of ammonia, a key ingredient in fertilizer. But as important as these other applications are to society—and they are very important—currently most natural gas is burned to produce either heat or power. To understand the value of natural gas to power producers, it is important to understand in some detail how the chemical energy stored in natural gas is converted into electrical energy. That is the goal of this chapter.

THE PRODUCTION OF HEAT

Natural gas, as it comes out of the pipeline system, is mostly methane, and for most of this chapter it will be modeled as if it were pure

Hartford, Connecticut, witnessed a power plant building boom after its power market was restructured. All of the new plants were natural gas–fired units. *(Hartford Tourism Council)*

methane. This is a helpful but simplifying assumption—that is, it is both false and true enough to be useful. Assuming natural gas to be pure methane makes the discussion simpler without leading the exposition too far from the truth.

Methane, an invisible and odorless gas, is a very simple hydrocarbon. A methane molecule is comprised of one carbon atom (chemical symbol C) bonded with four hydrogen atoms (chemical symbol H). The chemical symbol for methane is, therefore, CH_4. Combustion occurs when the carbon-hydrogen bonds of the CH_4 molecule are severed, at which point the carbon atoms bond with oxygen atoms, and the hydrogen atoms form bonds with still other oxygen atoms. The result is carbon dioxide (CO_2), water (H_2O), and heat. To be clear: A CO_2 molecule is formed when the carbon atom from a CH_4 molecule combines with two oxygen atoms, and the H_2O molecule is formed when two hydrogen atoms from a CH_4 combine with one oxygen atom. Provided that four oxygen atoms are available, each CH_4 molecule will, therefore, produce one molecule of CO_2 and two molecules of H_2O.

As a practical matter, the oxygen in the reaction comes from air, and the oxygen that exists in the air is in the form of oxygen molecules, two atoms of oxygen bound together. This is reflected in the chemical formula for molecular oxygen, which is O_2. The chemical equation for combustion of methane is, therefore,

$$CH_4 + 2O_2 \rightarrow CO_2 + 2H_2O$$

This equation does more than identify the reactants, the term for the molecules on the left side of the arrow, and the products, which is the term for the molecules on the right side of the arrow. It also indicates the ratios in which they occur. With respect to the reactants, the oxygen molecule to methane molecule ratio is 2:1. With respect to the products, the ratio of water molecules to carbon dioxide molecules is also 2:1. A combustion reaction that produces only carbon dioxide and water is "clean," in the sense that CO_2 and H_2O are the most benign products that can result from the burning of any hydrocarbon. Natural gas is the cleanest burning of all fossil fuels.

Even the combustion of natural gas can, in practice, produce other, more harmful, chemicals. If, for example, there is a deficit of oxygen atoms, carbon monoxide (CO), a highly poisonous gas, will be produced. And because natural gas is not pure methane, small amounts of other materials among the reactants lead to small amounts of other pollutants among the products. Engineers work hard to design maximally efficient combustion processes. In particular, the emissions produced by a natural gas–fired plant are almost exclusively water and carbon dioxide—as close as the best engineering practices will allow.

But the purpose of burning methane is to produce heat; the carbon dioxide and water are merely by-products. The production of heat can be understood as a consequence of the conservation of energy. It is a fundamental principle of science that energy cannot be created or destroyed. A certain amount of energy is required to bind the atoms of the reactant molecules together—that is, energy is required to bind the hydrogen and carbon atoms together to form the CH_4 molecule—and a certain amount of energy is required to bind the oxygen atoms together in the O_2 molecule. Energy is also required to form the product molecules, but *less* energy is required to form the CO_2 and the H_2O molecules than was required to form the reactant molecules—that is, the total chemical energy in the products is less than that in the reactants. As a consequence, there is energy "left over" after the product molecules have been formed. The difference in energy levels between the reactants and the products appears as heat. The production of heat in the combustion of methane is, therefore, often represented informally in the following way:

$$CH_4 + 2O_2 \rightarrow CO_2 + 2H_2O + heat$$

Expressing the amount of heat produced by each individual chemical reaction is possible but less informative than specifying the amount of heat produced by burning natural gas in bulk. As described in chapter 3, the amount of thermal energy produced by the combustion of bulk volumes of methane is expressed in terms of the heating value(s) of methane. The lower heating value (LHV) of methane is as 930 Btu per standard cubic foot (34.6 MJ/m³). Converting as much of this thermal energy into electrical energy as possible is the job of the generating station. The higher the percentage of thermal energy converted, the more efficient the conversion process.

ENERGY CONVERSION

Power plants are energy conversions devices—that is, they convert energy from one form to another—and most electricity is produced

General Electric F-series gas turbine *(GE Energy)*

by power plants that convert thermal energy into electrical energy. Examples of this type of conversion technology are natural gas–fired power plants, coal-fired power plants, and nuclear plants. Together these three power production technologies are responsible for almost 90 percent of the power produced in the United States. Natural gas–fired plants produce about 20 percent of electricity used in the United States. With few exceptions, most large national economies produce most of their electricity using some combination of these three technologies.

To convert thermal energy into electrical energy, these plants require the following four components:

> ➤ a heat source,
> ➤ a so-called working fluid, such as water,
> ➤ a turbine, which is used to convert the straight-line motion of the working fluid into rotary motion, and finally,
> ➤ a generator, which converts the rotary motion of the turbine into electricity.

The simplest natural gas–fired plants work in the following way: Gas is burned in a combustion chamber to produce heat. The heat generated by combustion is transferred to liquid water through a radiator-like device called a heat exchanger, which permits the transfer of heat but not mass between the hot combustion gases and water. Upon absorbing the heat, the liquid water turns to high-pressure steam. The steam flows away from the heat exchanger and toward the steam turbine. It leaves a high-pressure steam line through a valve that directs the rapidly expanding steam against the blades of the steam turbine, forcing the turbine to spin. The spinning shaft of the steam turbine drives the generator, which produces the electricity. Once past the steam turbine, the steam enters a device called a condenser, where the steam is cooled in order to convert it back to liquid water. The water is pumped back toward the combustion chamber, where the cycle begins again.

Modern natural gas plants often use a modified version of this design called combined cycle gas turbine (CCGT). CCGT technology keeps the steam cycle described in the preceding paragraph but augments it with a device called a gas turbine. In a CCGT process, the fuel-air mixture is compressed and then ignited. As the combustion gases rapidly expand, they drive the gas turbine. The gas turbine converts some of the thermal energy of the expanding combustion gases into rotational energy. This rotational energy is converted into electrical energy by a generator. The conversion process cools the combustion gases somewhat, but they are still very hot—hot enough to boil water—and so they are directed toward a heat exchanger, which transfers some of the thermal energy of the combustion gases to liquid water, which turns to steam. This is the start of the steam cycle described earlier. Combined cycle plants are more expensive to build than simpler steam cycle plants, but they make better use of the heat in the sense that they convert more of the thermal energy into electricity. They are, consequently, more efficient. But how efficient can this type or any other type of process be? How much of the thermal energy produced by burning natural gas can be converted into electricity?

For a long time, efficiency was not much of a concern for power producers. Natural gas was so inexpensive that a plant that converted between 35 and 40 percent of its thermal energy into electricity was considered good enough. The rest of the thermal energy was wasted—much of it went up the smokestack—but it was more cost-effective to buy additional gas than it was to find ways of burning it more efficiently. In the late 1990s, however, the price of natural gas began to rise sharply. At the end of 2005, for example, the price of natural gas was more than double what it had been at the end of 2002. Suddenly, the cost of fuel became a major concern. (It remains a major concern.) Older natural gas plants became very expensive to operate, and power producers increasingly found that the electricity produced with these plants was too expensive to sell on the open market. In response, power producers became interested in more efficient designs. CCGT technology is, for many, the technology of choice, because it can operate in the 50–60 percent range—that is, these plants typically produce roughly 50 percent more electricity than older designs for the same amount of fuel.

It might seem as if a particularly good design would convert all of the thermal energy produced by burning natural gas into electricity, but there are physical constraints on how much thermal energy can be converted into electrical energy, constraints that place strict limits on the ultimate efficiency of any plant. This is the problem: A natural gas–fired power plant will burn 100 percent of the gas; it will produce 100 percent of the thermal energy and 100 percent of the pollution; but it will only convert somewhat more than half of the thermal energy into electricity, and only modest improvements on this are possible. In particular, it is impossible to convert 100 percent of the thermal energy into electrical energy. To make better use of the energy that is released by combustion, some power plants
(continues on page 68)

Alternatives to Natural Gas

The value of natural gas as a fuel for generating electricity cannot be determined without comparing it to the alternatives. While, in principle, many different technologies are available for generating electricity, in practice the choices are more limited. Roughly 95 percent of the electricity sold in the United States is generated using one of four sources. While the precise percentages vary from year to year, the variation is not great, and large-scale change, because it is both expensive and difficult to implement, occurs only slowly. Each year natural gas–fired plants generate about 19–20 percent of the total power; coal generates roughly 49–50 percent; nuclear power generates about 19–20 percent; and hydroelectric generates about 7 percent. Oil is still used to generate 1–2 percent of the nation's electricity, and everything else—wind power, geothermal power, solar power, and so forth—is responsible for what little remains of the total. In theory, of course, some of these alternatives could be scaled up to replace a significant portion of the electricity currently produced by natural gas–fired plants, but this would be a huge and costly undertaking and will not happen quickly if it happens at all. (Owners of natural gas–fired generating stations, having paid for their plants, want to operate them in order to recover their investment and maximize their profit. In addition, siting and building replacement facilities are always expensive undertakings, and if a proposed generating station is not located near a high voltage transmission corridor, it can take many years to acquire the permits needed to build the corridor connecting the generating station to the grid, and additional years to fight the inevitable lawsuits by landowners and others who will strive to prevent the construction of a corridor across their properties or through public lands.) The following are how the major power-generation technologies compare with natural gas:

- Coal: The cost of coal, when priced according to dollars per unit of thermal energy produced, is much less expensive than natural gas and is not subject to the large price fluctuations that have plagued consumers of natural gas in the recent past. Whenever

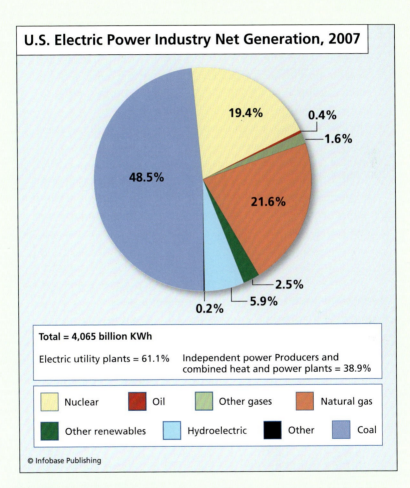

U.S. Electric Power Industry Net Generation, 2007

19.4%

0.4%

—1.6%

48.5%

21.6%

—2.5%

5.9%

0.2%

Total = 4,065 billion KWh

Electric utility plants = 61.1%   Independent power Producers and
combined heat and power plants = 38.9%

Nuclear   Oil   Other gases   Natural gas

Other renewables   Hydroelectric   Other   Coal

© Infobase Publishing

In 2006, approximately 95 percent of all electricity in the United States came from four sources, a situation that is unlikely to change much over the next decade. *(EIA)*

(continues)

(continued)

natural gas is used to provide base load power, coal-fired power plants can be used instead. Functionally, the big differences between coal and natural gas are (1) coal-fired power plants are less suitable for providing peak power because they take longer to start and stop than natural gas plants, and (2) burning coal produces far more power plant emissions of all types.

- Nuclear: The price of uranium, the key ingredient in nuclear fuel, is so small a part of generating electricity with nuclear power that electric rates of nuclear plants are almost completely insensitive to large fluctuations in the price of fuel. Consequently, the price of nuclear-generated electricity is extremely stable, enabling consumers to reliably make long-term plans. In contrast to fossil fuel plants, nuclear power plants produce no emissions. As with coal, nuclear plants work best when used to provide base load power and are not well suited for the sorts of hour-by-hour starts and stops characteristic of peak load power producers, an application for which natural gas plants are well suited.

- Hydroelectric: Hydroelectric power plants have the same flexibility that natural gas–fired plants exhibit: They work well for providing either peak or base load power. In fact, they are even more responsive to quick changes in demand than natural gas plants since power output of a hydroelectric facility rises or falls almost as soon as gates are opened or closed. And hydroelectric plants

(continued from page 65)

will use the thermal energy that they did not convert into work to heat a building. (While it is a basic principle of physics that not all the thermal energy can be converted into work, it can, for example, still be used for heating.) But with respect to the manufacture of electric power, in order to generate 1,000 units of electricity, the often produce zero emissions. They would be a good substitute for natural gas–fired plants except that in most nations all of the sites that are suitable for large-scale hydroelectric development have already been developed. In the United States, for example, the contribution that hydroelectric power makes to the total electricity supply steadily shrinks, not because the hydroelectric segment is generating less power, but because the total amount of power produced in the United States increases over time, and the contribution that hydroelectric power makes cannot be further increased.

Since the late 1990s, as a result of restructuring of the nation's electricity markets, most of the United States' growing demand for electricity has been met with natural gas–fired power plants. Today, these plants make a more important contribution to the nation's generating capacity than ever before. But because natural gas supplies and prices are not stable, increased reliance on these plants has increased volatility in the price of electricity. The predictability of the system has diminished. And as fuel prices have increased sharply, natural gas–fired plants have become less competitive with coal and nuclear power plants as base load power producers. It is expected that as the price of natural gas continues to rise, the rate of construction of new plants will continue to decrease. It remains to be seen what technologies will make up the difference.

power producer must, even with a modern well-maintained CCGT, burn sufficient natural gas to generate roughly 1,700–2,000 units of thermal energy. To convey what this means on a national scale, consider that in 2007 the United States burned 6.84 trillion cubic feet (192 billion cubic meters) of natural gas in the production of electricity. This was in addition to 4.72 trillion cubic feet (132 bil-

lion cubic meters) burned by residential consumers, and 9.64 trillion cubic feet (270 billion cubic meters) burned by commercial and industrial users. Such high levels of consumption cannot help but have significant environmental consequences.

EFFECTS ON THE ENVIRONMENT

The combustion of natural gas is not an end in itself but a means to an end. No one wants to expend energy per se. Instead, people want what the consumption of energy makes possible: electric lights, televisions, computers, air-conditioning, and so on. In many locations around the globe, the energy needed to power these devices is obtained by burning natural gas, and during the next few decades, many people will not be able to use these devices without consuming natural gas. Similar statements hold for the other major

Of all fossil fuel plants, natural gas plants such as the one pictured have the fewest harmful emissions. *(NOAA)*

natural gas markets: residential heating, commercial heating, and industrial applications.

But demands for electricity cause a chain of other activities to occur—including the exploration, production, transport, storage, and combustion of natural gas—and each link in this chain of activities has certain environmental consequences. Even under the best of circumstances, the environmental consequences associated with the natural gas business must be substantial because the scale of the business is enormous. Exploration and production are under way around the globe; hundreds of thousands of miles of pipeline are in continuous operation; trillions of cubic feet of natural gas are added and withdrawn from storage each year; and tens of trillions of cubic feet of natural gas are consumed each year. Are the environmental consequences excessive?

The question of whether natural gas is a good deal for consumers or for the environment has no meaning if it is not placed in context. Few people are willing to live without electricity, heat, and employment, and consequently, few people are willing to seriously entertain the notion of simply shutting off natural gas pipelines. An abrupt end to humanity's dependence on natural gas would cause economic chaos. In order to realistically evaluate the environmental consequences of natural gas as a primary energy source, it helps to compare them with those of the alternatives—coal and nuclear energy—the only two energy sources that could, over the next decade (at least), substitute for natural gas in terms of providing large amounts of baseload power. In what follows brief comparisons are made with coal and nuclear, and the reader is urged to think about the consequences of increased reliance on natural gas alternatives while reading about some of the environmental consequences of humankind's heavy reliance on natural gas. (Additional functional comparisons are found in the sidebar "Alternatives to Natural Gas.")

The most serious environmental consequences associated with the use of natural gas result from the combustion process. One way of understanding these consequences begins by describing natural gas in terms of its energy content and then to ask how much pollution is created for each million Btus of thermal energy produced. (A million Btus may appear to be large unit of measurement, but it is the amount of energy obtained by burning slightly less than 1,000 standard cubic feet [28 m^3] of natural gas. Alternatively, it is the amount of energy released in burning about nine gallons [44 l] of gasoline.)

When attempting to quantify the amount of pollution generated by burning natural gas in a practical combustion system, the equation shown on page 62, an idealized description of the combustion of natural gas, is a poor guide. The actual combustion of natural gas differs from the combustion reaction represented in the equation in three important ways. First, natural gas is not entirely methane, although methane is by far its largest single constituent; second, nitrogen is present during the combustion process because air, which is primarily nitrogen (not oxygen), is present in the combustion chamber; and third, methane is usually not completely combusted in a practical combustion system, resulting in the production of carbon monoxide.

According to the Energy Information Administration (EIA), for each million Btus (1,047 MJ) of energy obtained by burning natural gas:

> ➢ 117 pounds (53.1 kg) of CO_2 are produced. By way of comparison, generating a million Btus by burning coal will generate almost twice as much CO_2.

> ➢ 0.092 pounds (0.042 kg) of nitrogen oxides are produced. Nitrogen oxides are an important contributor to acid rain. Generating a million Btus by burning coal will generate almost five times as large a quantity of nitrogen oxides.

> ➢ 0.040 pounds (.018 kg) of carbon monoxide are produced. Generating the same amount of thermal energy

by burning coal will generate about five times as much carbon monoxide.

> ➤ 0.001 pounds (0.0005 kg) of sulfur dioxide, an impor-
> tant contributor to acid rain, are produced. Generating
> the same amount of energy by burning coal will gener-
> ate thousands of times as much sulfur dioxide.

In the United States, where more than 20 trillion cubic feet of natural gas are consumed each year, these figures translate into more than 1 billion *short tons* (900 million metric tons) of CO_2 per year, more than 920,000 short tons (830,000 metric tons) of nitrogen oxides, more than 40,000 short tons (36,000 metric tons) of carbon monoxide, and more than 10,000 short tons (9,000 metric tons) of sulfur dioxide. It is also worth noting that approximately 2.5 percent of all natural gas burned in the United States is burned by the pipeline system just to obtain the energy necessary to operate the pipeline system.

In addition to pollutants created by burning natural gas, there is also an important greenhouse gas released—unburned natural gas. In fact, methane (CH_4) is a much more potent greenhouse gas than CO_2—approximately 20 times as potent. This fact is significant, because natural gas is released during its production, transport, and storage. In order to facilitate comparisons, the amount of CH_4 released into the atmosphere is often expressed in terms of "CO_2 equivalent," the amount of CO_2 that would have to be released into the atmosphere to have the same effect as the released methane. The EIA estimates that the natural gas industry releases enough natural gas to have the same effect on the environment as 154 million short tons (140 metric tons) of CO_2. Much of this is released incidentally during the production, transport, and storage of natural gas.

It is tempting to say that these figures describe large releases of pollutants, but "large" in comparison to what? Coal, the most widely used fossil fuel, yields far more emissions per unit of thermal energy produced. These comparisons are important because debates about energy are not really debates about whether or not to use energy. Historically, sufficient energy has always, whenever possible, been produced to meet the demands of consumers regardless of the environmental costs. Today, few with access to energy have shown a willingness to live without it. The issue, then, is not *whether* the necessary energy should be produced but *how* it should be produced. In that context, natural gas is an environmental bargain relative to coal because the emissions associated with the use of natural gas are small in comparison to those produced by the same amount of coal, where coal and natural gas are measured in terms of their energy content. Of course, in contrast to nuclear power, natural gas emissions are extremely high because nuclear power plants are essentially emissions-free.

Information about the amount of CO_2 and CH_4 released into the atmosphere is important because of the way that these gases affect average global temperatures. This is the so-called greenhouse effect. To appreciate the greenhouse/Earth analogy, consider, first, what happens in an actual greenhouse. When the Sun shines on a greenhouse, the Sun's rays pass through its glass walls and roof because glass is transparent to light. Some of these rays strike the plants, tables, and other items inside the greenhouse and are reflected back through the glass, but some of these rays are absorbed by the plants and equipment inside the greenhouse and radiated back into the air as heat. From one point of view, light and heat are simply different forms of energy. But from the point of view of someone inside the greenhouse, the big difference between light and heat is that light passes easily through glass, and heat does not. Consequently, as light shines through the glass, heat accumulates inside the greenhouse and temperatures rise. The maximum temperature inside the greenhouse depends, in part, on the amount of light entering it, and the chemical properties of the glass of which it is made. Some types of glass are better at retaining heat than others.

On Earth, the atmosphere functions in a way that is similar to the glass in the greenhouse. Light from the Sun passes through the atmosphere, which is largely transparent to the Sun's rays, and strikes Earth's surface. Some of this light is reflected, and some of it is absorbed at the surface and radiated back into the atmosphere as heat. The atmosphere retains some of this heat. How much heat is retained depends on the chemical composition of the atmosphere. In particular, the higher the levels of CO_2 and CH_4 in the atmosphere, the more heat the atmosphere retains. This explains the analogy between Earth and a greenhouse and why increasing levels of CO_2 and CH_4 mean increased average global temperatures.

It is important to note that the atmosphere contained substantial amounts of CO_2 before humanity began to add CO_2 by burning fossil fuels. In fact, without some atmospheric CO_2, planetary temperatures would be very cold and Earth very hostile to life as it exists today. Moreover, levels of atmospheric CO_2 and CH_4 have previously fluctuated without human intervention, sometimes by large amounts. Scientists have discovered creative ways to document past changes in atmospheric chemistry and to correlate those changes with fluctuations in temperature, thereby adding to their understanding of global warming. What is different today is that humans are causing changes in atmospheric chemistry, principally through the burning of fossil fuels, and the rate at which humans are burning fossil fuels is accelerating.

While the analogy between Earth and a greenhouse is useful, it has its limitations. One of the most important has to do with energy transfer within the greenhouse. A greenhouse in the Sun is heated fairly evenly from one end to the other and temperature differences from floor to ceiling are modest and easily predictable. Earth, however, is a more complex system with a more complex geometry. In particular, Earth is heated very unevenly by the Sun, and this uneven heating causes the formation of large currents in the oceans and in the atmosphere that serve to distribute thermal

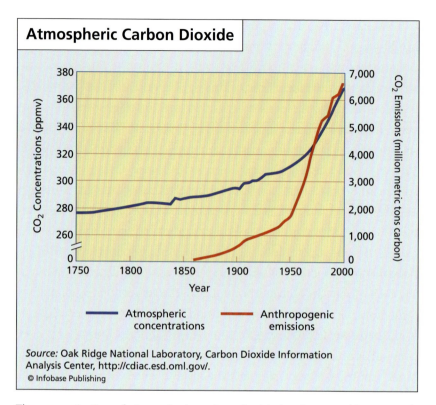

Atmospheric Carbon Dioxide

Source: Oak Ridge National Laboratory, Carbon Dioxide Information Analysis Center, http://cdiac.esd.oml.gov/.
© Infobase Publishing

The concentration of atmospheric carbon dioxide has increased in step with increased carbon dioxide emissions due to human activity. *(EIA)*

energy across the planet's surface. Some of these phenomena, such as the Gulf Stream and hurricanes, are well known and are frequently the subject of reports in the news media. Some, such as the thermohaline circulation, a planet-girdling deep ocean current, are less widely recognized. As the average temperatures of the atmosphere and oceans increase, some of these phenomena will increase in intensity and others will decrease. In particular, increasing ocean temperatures may eventually diminish the flow of the Gulf Stream, a current that carries warm tropical waters into the North Atlantic. If this enormous ocean current fails, temperatures may well fall in western Europe, because that region's mild winters are a result of the thermal energy carried northward by the Gulf Stream. So even as *average* temperatures increase, temperatures in some regions may decrease. Earth is no greenhouse.

Billions of people are now seeking the benefits of lifestyles that were once the almost exclusive province of the West. Historically, Western lifestyles have been energy-intensive—they still are—and most of that energy has come from fossil fuels. Reliable and reasonably priced electricity, comfortable homes, good health care, good education, a good transportation infrastructure, and rapid and reliable communications are some of the advantages associated with ready access to abundant and inexpensive energy. Developing nations are no more willing to put aside or delay their pursuit of these advantages than developed nations have been willing to abandon them in the name of the environment. The challenge, therefore, is to find ways to satisfy the aspirations of an ever-increasing number of people in ways that are not self-defeating. Whether this challenge can be met while remaining heavily dependent on fossil fuels is an open question.

The Restructured
Natural Gas Market

Economists, commentators, labor leaders, and business owners all spend time discussing "the" free market, but there are
many free markets. Each region or economic zone establishes rules
by which *market participants* operate, and although the rules differ somewhat from market to market, many of these markets can
be fairly described as free. As described in chapter 1, beginning in
the 1930s and continuing for several decades, the U.S. government
introduced increasingly intrusive regulations to protect consumers
from monopolistic practices that could have, in theory, been exer-
cised by some companies in the natural gas industry. The natural
gas market that evolved in response to these regulations was not
at all free nor did it provide consumers with adequate supplies of
natural gas. As an experiment, this type of regulation was a failure
in the sense that over time it led to results that fell far short of those
intended.

Power lines outside Argenta, Illinois. The demand for electricity continues to increase. *(Greg Munie)*

Beginning in the 1970s, the government introduced new laws and regulations with an eye toward restructuring the natural gas markets in order to make better use of free-market ideas. This is not to say that the natural gas market was deregulated. It was not. In particular, that sector of the natural gas industry concerned with the transportation of natural gas is still a "natural monopoly" in the sense that it is not possible—or at least not desirable—to allow multiple pipeline competitors to serve the same markets. To simply deregulate these monopolies would be irresponsible. Instead, the government sought a middle way, creating markets where competition was possible and strictly regulating those aspects of the business where it was not. The purpose of this chapter is to give some background on the natural gas markets as they currently exist.

PIPELINES: A NATURAL MONOPOLY

Throughout the history of the natural gas business, there have been numerous natural gas producers and numerous natural gas cus-

Natural gas pipeline construction. Almost all natural gas is transported by pipeline. *(BASF Coatings)*

tomers but relatively few pipeline companies. Moreover, pipeline companies operated without competition in the sense that only one pipeline connected a particular producing region with a particular market. Interstate pipeline operators were, therefore, able to control the entire market, since there is no other way to move large quantities of gas from place to place. (Pipeline operators that moved gas only within the state in which the producers were located were, for many years, in a slightly different situation, so here and in what follows the emphasis is on interstate pipeline companies.) The unique position of pipeline owners in natural gas markets has been recognized by all parties throughout the history of the natural gas industry.

For many years pipeline companies controlled all aspects of the natural gas industry either directly or indirectly. For decades, natural gas producers had to sell their gas to pipeline companies, not to the consumers that constituted their true market. The pipeline companies then transported their just-purchased gas either to market, where they sold it to local distribution companies, or to storage facilities. They also owned the storage facilities and all of their contents. To own the gas, to transport the gas, and to own and have sole access to the storage facilities is to have tremendous market power.

The federal government protected consumers by insisting on a certain price structure for natural gas. The idea was to assure consumers a fair price by limiting the power of the pipeline monopolies to charge unfair prices. But as mentioned in chapter 1, these price controls eventually created unintended shortages of natural gas, when producers became reluctant to find, produce, and sell natural gas to interstate pipeline operators because the price structure imposed by regulators had made profits impossible.

Another consequence of this highly regulated system were multiyear contracts between buyers and sellers in which buyers agreed to pay for gas shipments even if they later decided, for whatever reason, not to accept delivery of the gas. These contracts, called take-or-pay, resulted in prices for natural gas that bore little relationship to demand, a situation that guaranteed waste. Although established with the best of intentions, the natural gas regulatory system responsible for these peculiar business conditions was, by 1978, sorely in need of an overhaul.

As mentioned in chapter 1, the Natural Gas Policy Act (NGPA) of 1978 began a process of gradually deregulating natural gas prices. The NGPA was a useful beginning because it recognized the need for a new price structure, one that made the prospect of finding and developing additional gas supplies financially attractive to producers. But the NGPA was neither imaginative enough nor comprehensive enough to accomplish the goals it was designed to achieve. A bolder and more creative series of reforms began in 1985 when the Federal Energy Regulatory Commission (FERC) issued Order 436. This order gave pipeline companies the option of remaining in the business of purchasing and reselling the natural gas that flowed through their pipelines or of switching to a transportation-based business model, in which they charged customers for transporting natural gas but did not take ownership of the gas.

The option of acting solely as a conduit for the gas passing through their pipelines marked a fundamental change in the way some of the pipeline companies did business. Previously, the functions of ownership, transportation, and storage were bundled together and were perceived as different aspects of the same operation. Order 436 began a process by which the pipeline companies' multiple functions were unbundled. Competition was introduced wherever possible, and the pipeline companies' positions within the market were reduced and restricted as much as was practical. With pipeline companies restricted to transporting gas, producers and consumers were able to create new types of sales contracts. The new system opened up the possibility of price competition at each end of the pipeline and led to the development of a host of new and more efficient strategies for the buying and selling of natural gas. (See chapter 6.) Of course, in order for a more competitive model to work, pipeline companies, which still maintained their monopolies over the transport of gas, could not favor one company over another, and so the government required those companies that opted for this new business model to allocate pipeline capacity on a first come–first served basis.

The experiment begun by Order 436 was judged a success, and in 1992 FERC extended Order 436 by issuing Order 636, in which it created the domestic natural gas market in essentially the form in which it exists today. The option offered by FERC (and formulated in Order 436) to pipeline companies to grant to their customers open access to pipeline services was now made mandatory. Order 636 required pipeline companies to offer nondiscriminatory access to all market participants. The idea behind the order is that a pipeline should function as a sort of energy highway, offering access to all who want to use it, without imposing any unnecessary restrictions on the users. If a company had natural gas that met pipeline purity standards and wanted to transport its gas from point A to point B for whatever reason, the pipeline operator had to provide that company with access to the pipeline—provided, of course, that the company could pay the shipping cost. In the language of the act:

> The Commission must create a regulatory environment in which no gas seller has a competitive advantage over another gas seller. In particular, the Commission must regulate the pipeline transportation system and pipeline sales for resale in a manner that ensures that pipeline control of the transportation system—a natural monopoly—does not give a competitive advantage to pipelines over other sellers in the sale of natural gas.

Order 636 did not deregulate the natural gas industry as is sometimes claimed. The natural gas markets have never been deregulated. The charges that the pipeline companies could levy for the transport of natural gas, for example, are still regulated by FERC, and the requirement that each pipeline company grant nondiscriminatory access is strictly enforced. The difference between the new market and the old is that today's regulatory focus is much narrower than it was in the 1970s. After 1992, natural gas producers and consumers conducted business almost directly, but even now

(continues on page 86)

The Changing Nature of Natural Gas

Commercial natural gas is often described as almost pure methane. This is often true, but the chemical composition of this gas has begun to change to an increasingly complex mixture of gases, flammable and otherwise. There are three main reasons for the change:

1. the increasing price of natural gas,
2. increasing reliance on sources of natural gas–like fuels, and
3. the increasing demand for natural gas.

With respect to the first reason, natural gas, as it issues from a gas well, is often a mixture of methane and what are sometimes called natural gas liquids, the most familiar examples of which are propane, a common fuel for outdoor grills, and butane, a common fuel in lighters. Until the late 1990s, these hydrocarbons were separated prior to transporting the gas through the pipeline network. Natural gas liquids, despite their name, are commonly found in the gaseous state and can be burned together with methane, but for many years they were more valuable when sold separately. The petrochemical industry, for example, uses them in various industrial processes. When the price of natural gas began to rise sharply in the late 1990s, it became just as profitable to burn the natural gas liquids as it was to sell them separately to industry. Sometimes, depending on demand, natural gas liquids are now sold with methane as a fuel.

Second, some unconventional sources of gas have chemical compositions that differ substantially from the more conventional supplies. Coal-bed methane, for example, which is produced from coal deposits in Appalachia, contains significant amounts of nitrogen and carbon dioxide. Nitrogen and carbon dioxide contribute nothing to the combustion reaction, but they are not removed because it is uneconomical to remove them. Because of the presence of nitrogen and carbon dioxide, coal-bed methane has a significantly lower heating value than conventional natural gas—in other words, more coal-bed methane must be burned to produce the same amount of thermal energy as conventional natural gas. It is, in that sense, an inferior fuel. The market for coal-bed methane exists because producers occasionally encounter difficulty meeting the demand for conventional natural gas. But this introduces another difficulty: ensuring that all of these different potential fuels burn just as natural gas burns. Natural gas liquids, for example, generally have higher heating values than natural gas. To compensate for this difference, the gaseous fuels are mixed in order to provide consumers with predictable fuel performance.

Suppliers of natural gas increasingly speak about the interchangeability of fuels, by which they mean that although the chemical composition of the fuel flowing through the natural gas pipeline may change from hour to hour or season to season, the differences must remain invisible to the consumer. To accomplish interchangeability, suppliers must be sure that the fuel mixtures that they sell to consumers have roughly the same bulk characteristics as natural gas. The appliances and engines that have been designed to run on natural gas will not function properly unless the characteristics of the fuel are similar to those for which they were designed. The goal of natural gas suppliers is changing from one of providing consumers with almost pure methane to providing a fuel that retains the critical bulk characteristics of methane.

Whether gas supplies are interchangeable can be a matter of dispute. Owners of sophisticated and expensive technologies want gas with the narrowest possible variations in fuel characteristics, while suppliers seek more latitude in interchangeability standards in order to make better use of existing supplies. The final arbiter is FERC, which is bound by a policy that states, ". . . to the extent pipelines and their customers cannot resolve disputes over gas quality and interchangeability, those disputes can be brought before the Commission to be resolved on a case-by-case basis."

(continued from page 83)

pipeline operators are not quite neutral third parties. The analogy between a pipeline and a highway is not perfect. In order to maintain the integrity of a pipeline system, for example, a pipeline operator can require shippers to inject gas into the pipeline at times and places chosen by the operator. This is not an unrestricted free market, but the new regulations were designed to promote competition whenever possible. In particular, regulators attempted to create markets that would be more efficient in the sense that they would better reflect supply, demand, and the value of natural gas as a fuel source.

NATURAL GAS STORAGE: NO LONGER A MONOPOLY

The requirement that pipeline companies offer nondiscriminatory access to their pipelines was only one facet of the unbundling of pipeline services. (Recall that unbundling means that the suite of services that were once offered on a take-it-or-leave-it basis by the pipeline companies—natural gas sales, transport, and storage—were, as a matter of law, separated in such a way that ownership of the pipeline did not grant the pipeline operator an unfair advantage over competitors, sometimes called market participants, in non-pipeline aspects of the natural gas business.) Another important change in the natural gas markets as a result of FERC's Order 636 involved the very big and very important business of natural gas storage.

Trillions of cubic feet of natural gas are placed in storage and withdrawn from storage each year. Originally, these early storage facilities, most of which were depleted natural gas fields, were created and operated for the purpose of increasing system reliability during periods of fluctuating demand. Demand, for example, is higher on very cold days and lower on temperate days and natural gas supply must vary simultaneously with demand. During periods of higher

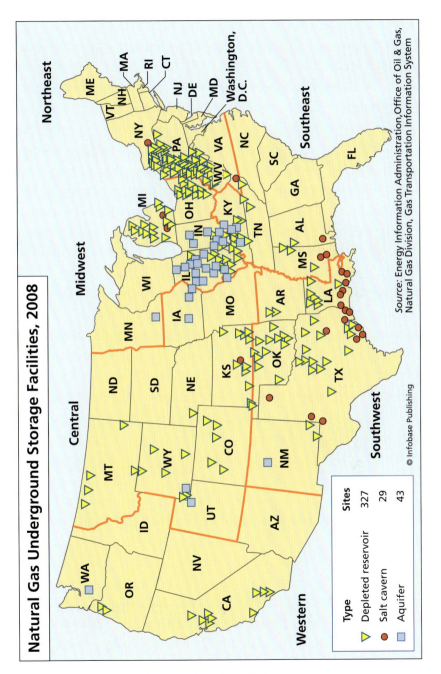

Natural Gas Underground Storage Facilities, 2008

Northeast

ME
VT NH MA
RI
CT
NY
PA
NJ
DE
MD
Washington, D.C.
VA
NC
WV
Southeast
SC
GA
FL

Midwest

MI
OH
KY
IN
TN
WI
IL
AL
MS
MO
AR
LA
IA
MN
ND
SD
NE
KS
OK
TX
CO
WY
NM
MT
UT
AZ
ID
NV
WA
OR
CA

Central

Southwest

Western

Source: Energy Information Administration, Office of Oil & Gas, Natural Gas Division, Gas Transportation Information System

© Infobase Publishing

Type	Sites
▷ Depleted reservoir	327
● Salt cavern	29
▣ Aquifer	43

Natural gas underground storage locations *(EIA)*

demand, natural gas producers may not be able to produce enough gas at that moment to enable the pipeline company to meet the usually brief spike in demand. Even if producers can, in theory, meet sharp increases in demand, a pipeline may not have sufficient capacity to transport the large volumes of natural gas involved. Therefore, prior to the restructuring of the natural gas markets, the pipeline companies created a system of strategically placed storage facilities that they could use to meet the demand for gas when demand was high. (These facilities are described in more detail in chapter 3.) The schedule by which the storage facilities were operated was seasonal. During the heating months, gas was, as a rule, withdrawn to meet demand, and during the warmer months gas was injected into the facilities for later use.

The storage facilities were a perfectly reasonable way of responding to fluctuations in demand, and the motives of the companies that owned the facilities and all of the gas within them were benign. But when the government began to consider the possibility of restructuring the natural gas markets, regulators recognized an opportunity to create a competitive business where none had existed. They created the natural gas storage business. To create these new opportunities, the function of storing natural gas had to be unbundled from those functions once performed solely by pipeline operators.

Interstate gas companies, which are the companies regulated by FERC, were directed to grant other companies—in the jargon of the industry they are often called market participants—equal access to their storage facilities so that any market participant that can afford the storage fees can use the storage facilities. Regulators recognized two important principles: First, pipeline companies had a legitimate need to reserve some storage space within their own storage facilities in order "balance the load" on the pipeline system—that is, they had to be able to remove or inject natural gas in order to keep supply in step with demand. They retain this right. Second, all market participants interested in using the surplus storage within the facility had to be granted nondiscriminatory access.

To see why a company might want to get into the natural gas storage business, consider an industry that uses a great deal of natural gas. Because natural gas purchases represent a large cost for such a business, the company's owners might decide to purchase their own supply of natural gas. The business could then reduce its overhead by buying gas when it was cheap, storing it, and then withdrawing the gas later when it was needed and when, presumably, the price would be higher. As long as the increase in the price of natural gas was greater than the storage fees charged to the business, such a strategy would enable the business to save money. It is, however, just a short step away from this practice to one of pure speculation: buying gas when the price is low and selling it when the price is high. Both types of business practices need to have access to natural gas storage facilities.

Today's natural gas storage business began in the facilities owned by the pipeline companies, most of which were depleted petroleum fields. But for certain aspects of the natural gas storage business, depleted fields are less than ideal. Each depleted natural gas (or oil) field has a maximum rate at which the gas can be stored or withdrawn. That rate depends on the geology of the field. Some fields are constituted of reservoir rocks that permit the gas to move more rapidly through the formation than others, but in the end all such fields have fairly severe limits on withdrawal rates.

For certain storage applications, operators want to store or withdraw gas very rapidly. The requirement for rapid transfers of gas into and out of storage together with the right to nondiscriminatory access to the pipeline network have made it possible for a number of business to earn a profit building and operating rapid-transfer storage sites. The best facilities for rapid storage and withdrawal are the salt caverns, which consist of huge voids created within enormous salt deposits. The amount of salt-cavern storage capacity has grown

Without adequate storage, seasonal fluctuations in demand due to residential heating could not be accommodated at a reasonable cost. *(Derick Jensen)*

rapidly since the restructuring of the natural gas markets. This is an example of a free market created (and maintained) by government regulation. (See chapter 3 for a more complete description of salt-cavern storage.)

The previous system of intensive regulation of all aspects of the natural gas markets was described as a failure because it produced an undesirable outcome, namely natural gas shortages. This raises the question of whether the current system has been a success. With respect to the natural gas storage business, the situation is complicated. On the one hand, some local gas distribution companies, which are not subject to FERC regulatory oversight, have of their own accord copied the example of the interstate pipeline companies and now maintain a minimum supply of gas within their own facilities for their own use and rent out the surplus capacity in an open and nondiscriminatory manner. That, together with the construction of salt caverns described in the preceding paragraph, indicates that the new model of the natural gas storage business is, at least, financially attractive.

But natural gas storage, which is now only partly concerned with load balancing, currently operates in ways that tend to decrease the predictability of the system. For example, consider the (simplified) case of a company that has natural gas in storage during a period in which prices are expected to increase. If the storage operator believes that by holding onto the gas it can earn a larger profit by selling later, it would be foolish not to do so. The purpose of the business is, after all, to maximize profits. Alternatively, the storage operator can offer to sell the gas immediately at the price that it anticipates being offered later. If enough operators decline to sell now except at a price that they expect to get later, a shortage or a price spike is created. Of course, as soon as the current price rises to the anticipated future price, gas inventories are released, and the "shortage" disappears. Natural gas storage, once used to make the system more predictable, now contributes to price instability as storage operators act in perfectly rational ways. (The actual situation is complicated by several factors, including the fact that storage operators must pay storage costs and possibly other costs for as long as the gas is in storage. So, for example, if the anticipated price increase is small, the small difference in selling price will be lost to the additional storage costs. Holding onto the gas under these circumstances is, therefore, a money-losing strategy.)

The supporters of the restructured natural gas market assert that the current way of doing business is more efficient in the sense that market participants are better able to respond to changes in information and other fluctuations in market conditions. And consumers are also market participants. In many areas, consumers can switch from one natural gas provider to another, and this, it is sometimes claimed, is a rational way of responding to price hikes on the part of their suppliers. But in practice, most consumers have neither the time nor the expertise to follow the *commodity* markets and respond "rationally." Instead, they pay the bill they receive. Whether consumers would not be better served by a differently designed market is an open question and one deserving more attention than it currently receives.

Energy Markets and Energy Politics

Natural gas has been bought and sold in the United States for more than a century, and during that time a great deal of thought has gone into finding the best way to "find" a price for natural gas that reflects its value. Early in the history of the industry, producers and consumers entered into individual contracts, examples of which have been described earlier in this volume, and hoped for the best. Beginning in the 1930s, recognizing that pipeline operators were in a unique position to prevent competition and dictate prices, the federal government became involved in setting natural gas prices. At first, regulatory efforts were modest and modestly successful in ensuring the goals of stable supplies, reasonable prices, and reasonable profits for producers. Over time, however, price controls were extended across an ever-increasing share of the natural gas industry, eventually producing a number of unintended and undesirable consequences.

World Financial Center, location of NYMEX, as seen from the Hudson River *(World Financial Center)*

By the 1970s, natural gas markets had become paralyzed by over-regulation and as a consequence no longer delivered enough natural gas to meet demand. In response, the federal government began to transition to a new system for buying and selling natural gas, one that operated under rules that were very different from the old ones. This new, restructured marketplace emphasized competition among "market participants," a general term for those engaged in the buying and selling of natural gas and related financial products. The idea was to create a marketplace that operated with less direct government intervention and in which many market participants competed openly and fairly with one another. It this market, it was believed, an optimal price for natural gas would develop—"optimal" in the sense that it best reflected the value of natural gas to producers and consumers alike. It was hoped that this new pricing scheme would provide producers with the incentives necessary to develop new sources of natural gas that would be sold at prices that consumers would find "just and reasonable."

The first goal of this chapter is to describe some of the basic features of the natural gas marketplace as it exists today. The second goal is to examine the role of government in assuring the supply of natural gas, a role that is clearly in the national interest.

PRICES AND PROFITS IN THE NATURAL GAS INDUSTRY

Beginning in the late 1970s, as federal price controls for natural gas were gradually withdrawn, natural gas buyers and sellers turned their attention to finding a mutually acceptable method of pricing natural gas. This may seem a simple problem; it has proved to not be simple at all.

The first thing to understand about the natural gas market is that natural gas is a commodity, by which is meant a material that is sold in bulk with characteristics that are uniform across shipments. Of course, the physical characteristics of natural gas as it issues from the wellhead vary substantially from one well to the next, but the bulk properties of the natural gas that is injected into a natural gas pipeline or a natural gas storage facility—its heating value, for example, and its moisture content—are standardized in the sense that they must lie within an acceptable range of variation. Natural gas is standardized near the wellhead in order to satisfy the quality requirements of the local pipeline operator, who will not accept for shipment any gas that does not meet its quality standards. Natural gas from one source should, therefore, be fully interchangeable with natural gas from any other source. In particular, natural gas producers cannot compete with one another on quality because they are all selling essentially the same thing. Unable to compete with respect to quality, they are left to compete on price.

This natural gas–fired unit, located within the ExxonMobil Complex in Baytown, Texas, produces both steam and electricity at an efficiency rate of about twice that of older steam cycle units. *(Exxon Mobil)*

The second thing to understand with respect to natural gas is that demand for this product fluctuates. Changes in the weather, changes in the season, and changes in the customer base all affect demand in different ways. Fluctuating demand is important to keep in mind because in the absence of price controls, changes in demand are reflected in changes in price. Natural gas prices also fluctuate in response to changes in the rate of production. Hurricanes along the Gulf Coast, for example, have occasionally interfered with natural gas production, causing prices to increase abruptly. Today, natural gas prices are very volatile. They are, in fact, some of the most volatile of all commodity prices.

There are many market participants in today's natural gas markets, and this is important because gas markets will not function efficiently unless many participants compete for a limited number of contracts. Some of these market participants are natural gas producers; some are local distribution companies, which are those businesses that deliver gas to end users; and there are many market participants who own neither production facilities nor distribution facilities. Some market participants, called traders, link buyers and sellers in a way that is at least a little similar to the way a grocer links farmers and families. Grocers buy and sell food, but they do not grow food themselves nor do they buy their food for personal consumption. And there are still other market participants who buy and sell contracts for the sale and purchase of natural gas with the understanding that their transactions will not result in the exchange of any natural gas. (More about these market participants, sometimes called financial traders, later.)

By way of example, suppose that a factory that uses a lot of natural gas expresses an interest in purchasing x units of natural gas for delivery in, say, three months, because it wants to assure itself of a future supply of fuel at a prearranged price. Traders, who buy and sell natural gas in the hope of earning short-term profits, will compete for the opportunity to win the contract to supply the gas. Many of the traders will not have the necessary gas when they make their bids, so their offers will reflect their best guesses about how much they must pay for the gas that they are promising to furnish as well as their estimates of any additional costs that they might incur—for example, storage, transport, and finance costs—and, of course, they must also include their profit margin. Traders who bid too low will have to purchase gas at a higher price than the price at which they agreed to sell it. Traders who bid too low will lose money; too many money-losing trades and they will be out of business. Some traders may already own the requisite volumes of gas, but they are also constrained not to bid too low. They will not, for example, offer their gas for a lower price than they could receive for it elsewhere. By contrast, traders who bid too high will not be awarded the sought-after contract. Too many failures and they, too, will soon be out of business.

This type of competitive system is designed to reward any innovations or economies that would enable a trader to bid lower than his or her competitors and still earn a profit. Competitors who do not at least occasionally distinguish themselves from their colleagues will eventually be driven from the marketplace. Nor are these "market forces" experienced by traders alone. It is the nature of the restructured natural gas markets that competitive pressures are distributed throughout the marketplace. In theory, market forces compel all market participants to continually search for ways to operate more efficiently. Those who find ways to increase their efficiency will be rewarded, and those who fail to adapt will be left behind.

Now consider the plight of the successful trader, the one who procured the contract to supply x units of natural gas in three months at a price of, say, y dollars. A major difficulty from the trader's viewpoint is that the price of gas to be purchased is not yet fixed even though the cost of the gas the trader has agreed to sell is now fixed by the sales contract. This could work to the trader's advantage. If the cash price of gas drops sharply before the delivery date, the trader can buy all of the needed gas at a price far below y dollars and earn a handsome profit. But the price could instead rise sharply, compelling the trader to spend far more than y dollars to buy the gas (in order to sell to the factory at y dollars). If the price rises in an unexpected way, the trader will lose money. Or the trader may have correctly guessed the future cash price of gas, in which case fulfillment of the contract will generate a modest profit. The problem is not that the trader cannot earn money from a contract to supply gas in the future at a price negotiated in the present but rather that the final profit depends on the future cash price of gas, a factor over which the trader can exercise no control. To ensure that the contract generates the profit first envisioned by the trader, the trader will turn to the natural gas *futures market,* a market that rarely entails the delivery of any gas at all.

The concept of a futures market was pioneered during the 19th century in Chicago, and the first futures markets involved agricultural commodities. As with today's natural gas traders, 19th-century traders in agricultural commodities found themselves agreeing to furnish agricultural commodities in the future at prices negotiated in the present. (Agricultural futures markets remain an important and vibrant part of agricultural markets.) A futures market—any futures market—is a public marketplace where commodities are contracted for purchase or sale.

The purchase and sales of *futures contracts* are handled by specialists called brokers, and brokers handle a great many futures contracts each day. The contracts are standardized in order to facilitate the purchase and sale of large numbers of them: The delivery location is, for example, standardized. In the case of natural gas futures contracts, the delivery location is the Henry Hub, which is located in Louisiana. Volumes of natural gas on the futures market are measured in standardized units, where one unit of gas is the amount needed to generate 10,000 million Btu (10,600 GJ), and 10,000 million, usually written 10,000 MMBtu, is the way that the amount is expressed in a gas futures contract. Price, of course, is open to negotiation. It is not unusual for more than 1 million of these contracts to be traded in a single day. Although delivery is always a possibility with a natural gas futures contract, less than 2 percent of these futures contracts result in the delivery of natural gas. Instead, it is the contracts themselves that are of interest to financial traders.

In just the same way that the cash natural gas market, which is a name for the market that is concerned with the delivery of natural gas, was structured so as to ensure as much competition as possible, the natural gas futures markets were also created to force competition between market participants. Futures markets are structured so that the prices of the contracts that are traded increase and decrease just as the price of natural gas increases and decreases. The cash price and futures price are seldom identical, but they generally do move in parallel—that is, as the price of one increases, so does the other. Similarly, when the price of one decreases the other will decrease as well and by roughly the same amount. In fact, if the price of natural gas futures ceased to move in parallel with the cash price of natural gas, the resulting price difference could be quickly converted into a profit by an alert trader. (There are several reasons that the two prices are seldom identical. For example, storing natural gas is a cost borne by the owner of the product, but "storing" a futures contract is free. The cost of storage would, therefore, be reflected in the cost of a cash contract but not a futures contract.)

The trader described earlier, the one who sold x units of natural gas for y dollars for delivery in three months, can protect the profit negotiated into the cash contract by simultaneously purchasing a futures contract to buy x units of natural gas for delivery in three months. The futures contract may specify a price of z dollars, and z may be higher, lower, or the same as y. From the trader's point of view, the actual price specified in the futures contract is not as important as the way that the price *changes* in the futures and *cash markets*. Again, the two prices move up and down in parallel.

Putting aside, for the moment, the fact that the futures contract will almost certainly not involve the actual delivery of natural gas, the procedure works like this: By agreeing to both buy and sell x units of natural gas for delivery in three months, any change in the price of gas will cause a loss on one contract and a gain on the other. Because the futures and cash prices move in parallel, the amount of loss on one contract is the same as the amount of profit on the other.

When the trader buys the natural gas needed to fulfill the contract—that is, when the trader makes a purchase on the cash market—he or she will simultaneously "offset" the futures contract, which is a procedure for selling the value of the gas that the trader agreed to buy on the futures market. The value of the contract is sold (offset), and the procedure relieves the trader of the responsibility of delivering physical quantities of gas because the amount the trader agreed to buy and the offset "balanced out." In effect, offsets are one method for "balancing the books." The practice of selling (or buying) on the cash market and simultaneously buying (or selling) on the futures market in this way is called *hedging.* The following is a summary of the hedge just described:

> Case #1: Assume that during the three-month interval between the time the contract is signed and the time the gas is to be delivered that the price of x units of natural gas increases on both the cash and futures markets by m dollars.

	CASH MARKET	FUTURES MARKET
initial actions	sell gas for y dollars	buy for z dollars
three months later	buy for $y + m$ dollars	sell (offset) for $z + m$ dollars
results	loss of m dollars	profit of m dollars

> Case #2: Assume that during the three-month interval between the time the contract is signed and the time the gas is to be delivered that the price of x units of natural gas decreases on both the cash and futures markets by m dollars.

	CASH MARKET	FUTURES MARKET
initial actions	sell gas for y dollars	buy for z dollars
three months later	buy for $y - m$ dollars	sell (offset) for $z - m$ dollars
results	profit of m dollars	loss of m dollars

The key point is that because of price parallelism between the cash and futures markets, a hedge helps to eliminate the effects of price fluctuations on profits. Of course, as Case #2 illustrates, the hedge also reduces the trader's ability to profit because of fluctuations in the price of gas. Losing the ability to profit from price decreases in order to protect one's profits from the effects of price increases in the cost of natural gas is a compromise that hedgers accept in order to reduce their vulnerability to price fluctuations that they cannot control or even foresee.

To be clear: The trader can still lose money on the cash contract. If the contract was unprofitable before the trader established the hedge it will be unprofitable afterward. The hedge is no substitute for a good contract. It is a method of reducing the risk of loss due to fluctuations in the future price of natural gas. (It might seem that the hedge eliminates rather than reduces the risk associated with price fluctuations, because the sum of the profit and loss on the two markets equals zero in both examples, but actual hedges are not quite as neat as this example indicates. A real hedge greatly reduces the trader's risk, but because of some of the finer points of hedging, which are not discussed here, it is possible to suffer a small loss or gain due to random price fluctuations even with a hedge.)

In North America natural gas futures contracts are traded on the New York Mercantile Exchange (NYMEX). It is, initially, a surprising fact that there is far more financial trading occurring on NYMEX than there is trading on the cash market throughout North America, but there are more ways of futures trading than the hedge-type transaction that was just described.

Participants in natural gas financial markets, the markets in which futures contracts are traded, are usually divided into two categories: those who use the markets to reduce risk and those who seek risk in order to profit by it. Hedgers belong to the first category; those in the second category are called *speculators*. The natural gas markets are rife with speculators, individuals whose goal is to profit from the existence of price variations. Speculation is often likened to gambling, and there are, to be sure, some similarities. Speculators seek to profit from the often-random price variations that occur in the natural gas markets just as gamblers seek to profit from random variations in the roll of the dice, the draw of the cards, or the outcome of a horse race. And just as the activities of gamblers contribute nothing to the outcome of the horse race, the activities of speculators do not result in the delivery of any natural gas. Speculators are only interested in the price fluctuations of natural gas; they have no interest in the commodity itself. This strikes some people as unproductive and objectionable.

Because of a general distaste for speculation on the part of some, there have, historically, been occasional attempts to curtail the role of speculators in commodity markets. These efforts have not succeeded because when speculators leave, they take their money with them. In just the same way that cash markets depend on a large number of traders competing openly and equally, futures markets also depend on large numbers of traders, whatever their motives, competing openly with one another and on an equal footing. Speculators are an integral part of this collection of market participants. More generally, speculation and speculators are an important part of any healthy futures market.

ENERGY POLICIES AND THEIR EFFECTS

One does not have to read newspapers very often or listen to news reports very carefully to hear references to *the* free market. Commentators often describe the free market as if it had an objective existence all its own—the way that gravity does. In this view, the free market operates best with a minimum of governmental regulation. Free markets are sometimes even described as straining to perform their vital functions beneath the oppressive weight of government regulations. Commentators with this view often describe the current

(continues on page 112)

An Interview with Ray Boswell about Methane Hydrate Research

Dr. Ray Boswell is a geologist by training and currently works at the U.S. Department of Energy's National Energy Technology Laboratory where he participates in, and manages, R&D programs in natural gas hydrates. Previously, he has managed programs in low-permeability sandstone formations, prepared regional natural gas resource assessments and worked as a development geologist in industry. He has served in leadership positions in gas hydrate field programs in the Indian Ocean, the Gulf of Mexico, and Alaska. The following interview took place on October 24, 2007, exclusively for this book.

Q: Dr. Boswell, the U.S. has very large deposits of methane hydrates within its territory. Japan has some. What are some of the other countries with large deposits of methane hydrates?

A: There is a small but steadily expanding group of countries that are actively pursuing gas hydrate research. For a lot of those, their focus is resource issues. But for others, such as those in the European Union, their research has been primarily driven by environmental concerns. Countries that have large resources? A lot of countries have not yet done the foundational research to establish resource volumes, but those countries that have looked for hydrates have found deposits of interest. China recently had a drilling core expedition. South Korea has one ongoing now [Fall 2007]. India did an expedition last year and found some very rich deposits.

Q: What are the geometries of hydrates? Do you usually find them in thin sheets as if they were laid down in sediments? As particulate? As large blocks?

A: Hydrate is something that forms within the sediment, typically well after the sediment has been deposited. You have sediment in the gas hydrate stability zone, and if there is gas and water available and the

Ray Boswell at work. On the left is Dr. Timothy Collett, researcher with the U.S. Geological Survey. *(Dr. Ray Boswell)*

geochemical conditions are right, then hydrate can form within the pore space of the sediment. There are a wide range of forms that it takes.

In the marine environment there are three primary modes of occurrence for hydrates. The first is pore-filling. Pore-filling gas hydrates in reservoir quality marine sandstone have been seen in Japan, in the Gulf of Mexico, and elsewhere. These are the reservoirs that have the porosity and permeability that give them the best potential to be producing reservoirs. They are the same sorts of reservoirs that produce conventional oil and gas, but the pore space is filled with hydrate. The hydrate is disseminated throughout the mass of rock—and the high permeability enables rich concentrations of methane. Typically 60 to 80 percent of the pore space is filled with hydrate.

(continues)

(continued)

In finer-grained sediments, typically the physics are such that you can't get saturations of pore-filling hydrate that are as high. These make up the bulk of gas hydrate occurrence in marine strata on a volume basis. One large deposit off the coast of North Carolina on the Blake Ridge is a classic example of such a deposit. There is a lot of hydrate over a large area, but it is finely disseminated—typically less than 10 percent of the available pore volume is filled with hydrate. In China recently, we've seen some of the highest concentrations of pore-filling hydrate in a fine-grained deposit, with values approaching 40 percent, but still significantly less than you can get in sand deposits.

The second mode is an interesting one: If you have a section that is structurally deformed—there is fracturing of the rock—then the permeability of the rock is enhanced in that way, and that allows for greater concentrations. In that situation you can get thick sections of fine-grain sediment that is very highly saturated in gas hydrate, and in those cases the gas hydrate not only fills the pores, but occurs in vertical or near vertical veins, as thin sheets between sediment layers, as nodules, and in other massive forms.

The third form is as massive mounds of solid gas hydrate, often sitting directly on the seafloor. There are very spectacular pictures of those sorts of things. They support unique chemosynthetic communities. There are very good examples of those in the Gulf of Mexico and offshore Vancouver Island.

Q: With respect to the "fractured" fine-grained deposits you described— these are sedimentary beds of rocks where the tops and bottoms of the beds are connected by vertical fractures, but the vertical fractures are not connected to one another by horizontal fractures?

A: I'm thinking of a zone where there are pervasive disturbances, lots of faults and lots of fractures, and, yes, most of those would be vertical. In India, we discovered one very good site of that type. There was a lot of gas hydrate in a thick section.

Q: When you say that it is a thick section, how thick is it?

A: More than 100 meters thick in India. Korea has also recently announced finding a similar occurrence.

Q: One hundred meters thick—and what would be the extent of this bed horizontally?

A: That is not well known. Further drilling will be needed to delineate such structures.

Q: People have pulled nodules up (to the surface) and set them on fire?

A: Fisherman have been known to scrape along the seafloor with their nets in various places and bring solid gas hydrates to the surface, yes. You also get fairly good-sized nodules in the fractured-disturbed sediment type that can survive the trip to the surface during scientific coring operations. You get a big solid lump that you can set on fire if you so desire.

Q: I've seen pictures of them set on fire. It's spectacular.

When drilling for petroleum, drill operators don't like to encounter methane hydrates, because the wall of the drill hole can collapse when the hydrate dissociates during drilling. With respect to the large deposits of gas hydrates, how stable are they?

A: Hydrates are stable in their pressure and temperature condition. They also have a self-correcting mechanism. Reduce the pressure and some hydrate dissociates, but because it's an endothermic reaction, it also reduces the temperature around it. This tends to drive the hydrate back to being stable again. There is that issue to take into account when considering hydrate stability. But, yes, you can dissociate hydrate by changing its temperature or its pressure. There is a lot of work going on right now to figure out how big an issue stability is—to know how far away from a wellbore heat can propagate to create this sediment weakening, for example. There are also large-scale global events that can make hydrate unstable over geologic time.

(continues)

(continued)

Q: There is some evidence off the coast of North Carolina of shallow troughs that are very large in extent that some think were caused by massive dissociation of hydrates. And there is some evidence that massive dissociation of hydrates has occurred in geologic time, events big enough to change the temperature of the planet. With respect to deposits of methane hydrates around the Arctic, how stable are the hydrates that exist in the ground? And how stable are the hydrates that exist in the ocean toward the edge of that hydrate zone of stability?

A: As far as the Blake Ridge (off the coast of North Carolina) goes, it is not necessarily hydrate dissociation in that region that has caused the features you describe. Continually, there is methane that percolates upward through that section. As it converts to hydrate, it can form a cap that hinders further gas migration. As gas builds up underneath, eventually it creates too much pressure, resulting in fracturing of the sediment. The gas can escape through these fractures, and if there is a large-scale venting of gas, it can disturb the sediment and create those features. That's not really gas hydrate dissociation, but it is a gas hydrate–related phenomenon. It's gas hydrate interrupting methane flux through the sediment, creating a "logjam" that is later broken.

The climate link is interesting because there are very compelling arguments about how hydrate can be linked to long-term climate change. There is a noted and clear correlation between past temperatures and past methane levels in the atmosphere, and it is unclear what natural process could create large-scale changes in methane levels in the atmosphere other than large-scale hydrate dissociation. There is a linkage there. However, recently, people have focused on the timing of the temperature increase and the record of atmospheric methane in ice cores and have found that the methane release may actually follow the temperature change and not cause it. So while it's clear that there is a link—that they are happening as some sort of process—it is not yet clear if the methane is coming out as a driver of the temperature change or if the methane is coming out as a result of the temperature change. This is a topic that is widely discussed. I don't think there is a definitive answer on that, but it is certainly an interesting and important topic to understand.

It is a very compelling story because you can imagine a scenario in which you have hydrates that are under pressure and stabilized. But during an ice age, for example, a lower sea level would mean a lower pressure at the seafloor, and that could result in hydrate dissociation on the seafloor. However, there is also counterbalancing water-cooling during glacial periods that would tend to increase hydrate stability. Overall, it appears that cooler periods may lead to a net increase in methane stored as hydrate globally. In warmer periods, warming ocean waters most likely lead to a net destabilization of gas hydrate, both in the Arctic, and in the deep seas, where warming bottom waters more than compensate for the stabilizing effect of increased pressure from higher sea levels.

In any case, there is much we do not know about how methane from destabilized gas hydrate might find its way to the atmosphere. There are a lot of things that might sequester or consume it on the way through both the sediment and the water column—but if you have enough methane at one time it does seem possible that some could get through the water column to the atmosphere and create a global warming effect.

But other people have looked at the timing of the temperature increase and the record of atmospheric methane in ice cores and have found that the methane release may actually follow the temperature change and not to cause it. So while it's clear that there is a link—that they are happening as some sort of process—it is not yet clear if the methane is coming out as a driver of the temperature change or if the methane is coming out as a result of the temperature change. Before or after? That is a topic that is widely discussed. I don't think there is a definitive answer on that, but it is certainly an interesting and important topic to understand.

One other thing to understand about methane hydrate is that the vast bulk of the methane hydrate that we are talking about exists well below the seafloor, hundreds of meters below the seafloor. Only a very

(continues)

(continued)

small percentage is at the seafloor. Temperature perturbations take a long time to get down there, and if they do get down there, methane that dissociates has a lot of sediment to travel through and then a lot of water to travel through. There are a lot of barriers for that methane before it gets out of the sediment, through the water column, and to the atmosphere, but it certainly is possible. There is no doubt that it is possible, but it is not a situation in which if the temperature changes, then large portions of the global hydrate reservoir are converted to methane and immediately released to the atmosphere.

Q: So the marine hydrates are very stable. What about the ones locked in the tundra?

A: I don't mean to say they are very stable, but they are insulated from a lot of the instability by being deeply buried. Now one issue about that is that you can have large-scale slumps of the continental shelf, which remove that overburden quickly and expose those hydrate-bearing sediments to the atmosphere—but even the biggest of those slumps does not have effects that are global in scale. And some of the biggest seafloor slumps, such as the Storegga slide off the coast of Norway that has commonly been linked to gas hydrate, now appears to be due to non-hydrate factors. However, its not clear yet that such large-scale slumping does not occur—it is a very complicated geologic issue. There is not a lot of data to inform the analysis.

Arctic hydrates are another beast entirely. They are primarily in coarser-grained sandstone reservoirs not encased in this largely impermeable mud. They are in porous reservoirs. There is not a whole lot known about them. They have been studied a little in Russia, extensively at one site in Canada, and also in the area around Prudhoe Bay (Alaska). We do suspect that the gas hydrates around Prudhoe Bay, for example, are in traps that were preexisting free gas traps, which is to say, just another typical gas reservoir that is a part of the greater Prudhoe Bay petroleum system, the same sorts of formations that produce the oil and gas at Prudhoe Bay. That gas just kept migrating up and these [hydrate formations] are the highest traps that trapped that gas. Then with time, that part of Alaska moved north, and hydrate stability conditions were imposed on it. Through some process that we don't understand, a lot of those free gas traps seem to have been converted into hydrate accumulations. It's a different sort of system than the marine hydrates—the way they form and also the types of reservoirs that they are in. There are various things about that that we still need to learn. So, if these are preexisting gas traps, dissociation of gas hydrate should recreate the free gas traps, which should not leak methane any more than any other shallow gas trap. But it's not a process that anyone has witnessed or recreated in the lab.

Clearly, there are issues.

And there is a lot of methane trapped as hydrate in the Arctic that may not be in these preexisting free traps but may have been converted into hydrate on its way to the surface. If the hydrates dissociate then the methane will continue its migration, and you will have gas that should be similar to what you get over other oil and gas fields. The issue is that it is a naturally slow process that is interrupted by the hydrate formation process. We really don't know yet at what rate and how much dissociated hydrate will leak if the conditions in the Arctic continue to change.

Q: With respect to commercial development of hydrates, I gather that the methods all entail converting the hydrate back to methane gas and then drawing it out. How does production technology research stand at the present time?

A: There are three general approaches to gas hydrate production. One is reducing the pressure; one is increasing the temperature; and the third is injecting chemical inhibitors that change the stability conditions—like throwing salt on ice on the sidewalk. Of the work that has been done, largely in computer modeling—which is where most of this work occurs because there aren't a lot of field tests—it's now thought that the pressurization techniques should be the most effective. Thermal stimulation with hot water was attempted by an international consortium involving Japan, Canada, the U.S., and others in 2002 in the Northwest Territories.

(continues)

(continued)

That program also conducted depressurization tests and both of those tests showed that you could convert hydrate back into methane gas in the subsurface and then pump it out to the surface and get production. They proved the technical feasibility. In 2007, there was a project on the North Slope of Alaska with DOE (Department of Energy), BP and the USGS (United States Geological Survey) that also used a depressurization method to demonstrate the producibility of hydrate. So most R & D efforts are starting from the assumption that depressurization is the most likely method. But it is very likely that it will need to be supplemented by some heat to combat this endothermic cooling effect that will be happening near the wellbore and eventually in the wellbore as hydrate dissociation causes the refreezing of water and perhaps reformation of the hydrate.

Q: When would you expect a commercial well to begin producing methane from hydrate, and where do you suppose that will occur?

A: A lot of international programs, including ours, are aiming at 2015 as a goal, but that goal may mean different things for different countries. For a lot of countries, they want to begin to produce gas from hydrate by that time. In the U.S., what we want to do is to have the production and the knowledge in place so that production can begin at that time if it is economically viable, if it is competitive. In some other countries, that is less of an issue. Countries like Japan and India are very reliant on external sources of energy and expect to have huge and increasing demands for energy. In the U.S. what we are trying to do is establish another resource option that is available to deal with contingencies, unexpected events, and disruptions in supply. So by 2015, we expect to know what we need to know about Arctic hydrates and will have demonstrated that they can be produced and to understand what

(continued from page 103)

natural gas markets as "deregulated," but this type of description makes it difficult to appreciate what has happened in the natural gas industry and why.

is needed to produce them so that they can be a viable option if conditions warrant. For the marine hydrates, such as in the Gulf of Mexico, our goal is to be at the same point by 2020. It depends on the intensity of the research. Japan is spending probably four to five times more on research than we do in the U.S., so they can have an earlier goal for their marine resource. That's probably one reason that they have an earlier expectation.

Q: How confident are you that this phenomenon will be an important contribution to the nation's energy security?

A: The DOE program was founded on the recognition that there is an awful lot of methane hydrate out there, and that means profound things for the way we understand the environment, global carbon cycling, as well as future energy resource availability for the country. It raises very serious issues that need to be addressed. What we're trying to do is find out as quickly as possible gas hydrate's role in environmental processes, and whether methane hydrate can be a viable and environmentally re-sponsible resource of natural gas. It exists at such a scale that it certainly seems that it has good potential, but there is a lot of work that needs to be done and the detailed data needed is only now arriving. The timelines are long—2015 and 2020—these are timelines that are beyond what the industry typically would be interested in investing money. That's why the federal government is involved. There is a large potential and a lot of uncertainty, and a lot of interdisciplinary work that needs to be done in order to understand the role and potential of gas hydrates.

Q: Thank you very much for sharing your insights.

A: You're welcome.

Today's restructured natural gas markets have a key feature in common with the old pre-1978 natural gas industry. Federal legislation and regulation created both markets, and strict govern-mental oversight has been critical to the operation of both. This is important: The current natural gas markets are creations of the federal government just as the old markets were. Federal controls give expression to the view that the natural gas industry is (and has long been) vital to the nation's prosperity and security. They also recognize the unusual structure of an industry that is extremely vulnerable to the actions of a few key market participants—namely, pipeline operators and storage facility operators. Pipelines and storage facilities are parts of the business that are as much a concern for today's legislators and regulators as they were for legislators and regulators in generations past.

Initially, with the passage of the 1938 Natural Gas Act, the government responded to concerns about natural monopolies in the natural gas industry by placing controls on the price of natural gas that was shipped across state lines. For many years, price controls worked as they were designed to work: They assured consumers of abundant supplies of natural gas at prices that were stable and affordable, and producers and other market participants simultaneously earned reasonable profits. But this was in the days of easy-to-produce natural gas. In fact, during the first few decades of price controls, enough gas was produced incidentally as a by-product of the production of domestic oil that only a fraction of the gas produced from domestic oil fields was needed to supply the nation's natural gas system. The remainder of the gas was useless; it was just flared—bled off at the well and ignited as a safety hazard. Flaring occurred on a scale that is hard for a modern reader to appreciate. (An excellent and brief account of natural gas flaring can be found in the writings of Alistair Cooke, the British-born American reporter and commentator. Cooke first saw Dallas, Texas, in 1933 at twilight from a distant hilltop and was astounded to see the entire horizon illuminated by roaring gas flares, continuously burning the then-valueless natural gas.)

As the natural gas market changed, some sources of natural gas began to cost more to develop, and the price of alternatives to natural gas became uncompetitive because the price controls then in effect had distorted the value of natural gas to consumers, keeping the price too low relative to the alternatives. The system of price controls begun in 1938 eventually collapsed, not because it had not done its original job (it had) but because it had not successfully adapted to changing market conditions. Those who had designed the system of price controls had failed to put in place a mechanism that would allow the market that they had created to successfully adapt to changing market conditions.

The legislation and accompanying regulations that have created the current natural gas markets have sought to create a marketplace that can more readily adapt to changing conditions. Those parts of the industry most susceptible to manipulation are still under tight government supervision, because they are still natural monopolies, but natural gas *prices* are no longer considered to be part of the government's responsibility. Instead, the government has attempted to create a market in which the price of gas reflects the value of gas to producer and consumer alike. In the restructured natural gas marketplace, prices represent information as well as profits. This is the big difference between the old markets and the new. Under the old system, prices remained the same no matter the supply or the demand. By allowing prices to move freely, prices are more apt to accurately reflect the balance between supply and demand. Market participants can, therefore, better use prices to guide their investment decisions.

In order that prices represent information about the state of the market, the markets must be operated in a way that is both transparent and competitive. The role of competition in the natural gas markets was described in the preceding section. Transparency—that is, the free and open flow of information—is also an important aspect of the natural gas markets. A great deal of real-time or near real-time information about the price of natural gas, the price of natural gas futures, as well as many other useful statistics is published by the federal government, by NYMEX, and by several other authoritative sources. Price "signals," together with a transparent, fair, and competitive market, enable market participants to more easily identify methods of investing in the natural gas industry in ways that will be more profitable and hopefully more productive. The markets perform this way by design. This competitive model is used only in those areas of the market where the designers of the market thought that competition would be both possible and productive. Where competition seemed unlikely or undesirable—the pipeline segment of the industry, for example—strict government regulations were put in place that severely constrain how those participants can manage their own assets.

The rules under which today's natural gas markets operate could be changed. Some of the choices made by the designers of today's market were, at least to some extent, arbitrary, and different decisions could have been made that would have led to a somewhat different natural gas marketplace—not necessarily more restrictive or less restrictive decisions, but simply different ones.

It is doubtful that the current system is optimal. As a general rule, there are few legislative or regulatory efforts that, in retrospect, could not have been improved. That is probably also the case with the restructured natural gas markets. To improve the current system, one must, however, be clear about its purpose. Is the primary purpose of a successful marketplace to provide a place where market participants can attempt to earn profits or is the purpose of the marketplace to provide consumers with access to a vital resource at an affordable and predictable price? Or should the success of the market be measured in another way? Discussions about how well the restructured markets are operating—and even how they should be evaluated—and how the functioning of the markets might be improved are ongoing. Because they affect investment decisions by both producers and end users of natural gas, these are some of the most interesting and important topics in the natural gas industry today.

PART II

Hydrogen

The Physical and Chemical Properties of Hydrogen

Hydrogen is often touted as the fuel of the future, and there are, to be sure, a number of advantages to using hydrogen as a fuel. It burns easily, and it releases a great deal of heat per unit mass. (One unit mass of hydrogen releases approximately three times as much heat as one unit mass of gasoline.) Hydrogen can also be used to power energy conversion devices called fuel cells via a process that is more efficient than combustion. And hydrogen is, in theory, a renewable fuel in the sense that water is both a source of hydrogen and a product of its use. These advantages lead many to assert that it is only a matter of time before hydrogen becomes the fuel of choice, replacing fossil fuels and creating a new type of economy.

There are, however, serious barriers to the practical use of hydrogen as a fuel:

> ➤ It is expensive to produce.
> ➤ It is expensive to transport.

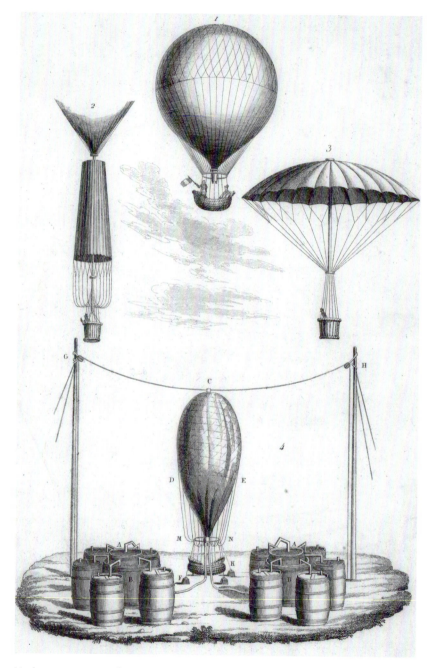

Hydrogen as an early transportation fuel *(Library of Congress Prints & Photographs Division)*

➢ As an automotive fuel, it is difficult to store on board in amounts large enough to provide drivers with a practical driving range.

➢ Fuel cells have so far proved to be expensive to manufacture and too prone to failure for use in many practical applications.

Each of these barriers has been the subject of intensive research for decades, and not one of them has been completely overcome. At this point, it is reasonable to ask whether satisfactory solutions exist for all of them—which is not to say that solutions do not exist but only to emphasize the importance of keeping an open mind on the subject.

The present chapter describes the physics of hydrogen, its production, transport, storage, and consumption, with special emphasis on what must be done to create a "hydrogen economy"—an economy where one or more fossil fuels are replaced by hydrogen. (Oil is the most important fossil fuel in this respect, because it is the only fossil fuel for which, even in theory, there currently exists no readily available substitute.)

HYDROGEN PRODUCTION

Hydrogen is the third most abundant element near Earth's surface, but on Earth hydrogen atoms are seldom found except in combination with other elements in such compounds as, for example, methane (CH_4), water (H_2O), and ammonia (NH_3). The absence of significant quantities of free hydrogen anywhere on Earth has profound consequences for its value as a fuel.

The first and most basic fact about hydrogen as a fuel is that it is a secondary energy source. Hydrogen has more in common with electricity, which is also created through the consumption of primary energy sources, than it does with natural gas, coal, oil, and uranium, which are primary sources of energy. The difference is that primary energy sources are there for the taking. They may need to be processed prior to consumption. Natural gas, for example, is usually processed before it is burned. Oil must be refined before it can be used in the transportation sector, and uranium must be processed to remove impurities; sometimes it must also be enriched before it can be used as fuel in a commercial nuclear generating station. But the energy required for these processes can, in theory, be obtained from the fuels themselves. More energy is obtained from uranium fuel, for example, than is required to process uranium ore to produce the fuel. Primary energy sources are, in this sense, energy-rich. The situation is very different with hydrogen: More energy is expended manufacturing hydrogen than is released by its consumption. In other words, if one used x units of hydrogen to power a hydrogen manufacturing process, one would have less than x units of hydrogen at the completion of the process because more energy would have been spent manufacturing the new hydrogen than was present in the original hydrogen supply.

To see the problem of hydrogen production more clearly, consider the process of *electrolysis*, in which hydrogen gas is produced by passing an electric current through water in a process that is demonstrated in many high school chemistry classes. In electrolysis, more electrical energy is required to produce a given amount of hydrogen gas than is released when the resulting hydrogen is consumed as a fuel. As a consequence, it is not practical to produce hydrogen via electrolysis when hydrogen is used as a fuel to power the electrolysis process. To do so would mean that there would be less hydrogen available at the end of the process than there was at the beginning. Compare this energy balance with that of oil, which is used to power the tankers that transport it to the refineries and can be used to power the refineries themselves and can also be used to power the pipelines and trucks that distribute refined petroleum to end users. Despite all of these energy "expenditures," there is still enough refined petroleum remaining to power the cars, trucks, trains, planes, and ships that use it for fuel for other activities. This energy difference is one reason why petroleum and not hydrogen powers the world's economies today.

But the process of electrolysis described in the preceding paragraph was only chosen for its familiarity. There are other ways of producing hydrogen from water, and there are other hydrogen sources, also called feedstocks, that can be used instead of water. The fact remains, however, that no matter the method of hydrogen production one chooses and no matter the feedstock one employs, hydrogen gas requires more energy to produce than one can recover by using it as fuel. To produce abundant quantities of hydrogen, therefore, one must have access to an abundant and inexpensive primary energy source to power the manufacturing process as well as a practical feedstock. The first barrier to replacing fossil fuels with hydrogen is, therefore, the difficulty of manufacturing large amounts of hydrogen gas in an economical way.

Currently, the principal hydrogen feedstock is methane (chemical symbol: CH_4). Hydrogen is produced from methane by a process called *reforming,* which involves separating hydrogen atoms from the carbon atoms to which they are bound. The result is hydrogen gas, which consists of hydrogen molecules formed from two hydrogen atoms—the symbol for molecular hydrogen is H_2—as well as a good deal of carbon dioxide (CO_2), the principal greenhouse gas.

There are two important advantages to reforming methane to produce hydrogen. First, natural gas is a hydrogen-rich fuel; each molecule of methane has four hydrogen atoms per carbon atom. Second, the technology used to reform methane is reliable and well understood. But there are also two significant disadvantages to using natural gas. First, the process generates a significant amount of CO_2 as a by-product, and this CO_2 is vented to the atmosphere. Second, in the United States, there is no surplus of methane. As mentioned in the first section of this book, exactly the opposite situation prevails. Domestic production of natural gas has not kept pace with demand, and using natural gas to produce hydrogen in sufficient quantities to replace fossil fuels would require enormous amounts of natural gas above what is currently used. In the absence of a successful methane hydrates program, natural gas imports and prices would soar. Energy security would not be enhanced by using natural gas as the principal feedstock for a hydrogen economy. Natural gas is, therefore, not currently a practical feedstock for large-scale hydrogen generation.

Coal can also be used to produce hydrogen. There are several advantages to using coal. First, coal is inexpensive and likely to remain so for the indefinite future. Second, coal can be used as a feedstock by employing technology that is already fairly well developed. Third, there is enough coal to serve as a primary energy source for a hydrogen economy for centuries, and in the United States that coal is contained within the national boundaries. As a consequence, the use of coal as a feedstock and as the primary energy source for the manufacture of hydrogen increases the energy security of the nation. The chief disadvantage to the use of coal is that even more CO_2 is produced per unit mass of hydrogen output that is the case when natural gas is used. In fact, manufacturing hydrogen from coal produces more CO_2 than any other method of hydrogen production. It is, as described in the first half of this volume, possible to sequester CO_2, but this drives up both the costs and the complexities of using coal as a feedstock, and a sequestration infrastructure is currently not in place. In fact the technology is not entirely proven.

One can also produce hydrogen from plant matter, also called *biomass*, such as switchgrass and corn, but probably not in the quantities necessary to power a hydrogen economy. The production of hydrogen from plant matter is too expensive and energy intensive: Fields must be fertilized and harvested, and the plant matter must be transported and processed. (The best current technology involves converting the biomass into a gas that would be reformed to produce the hydrogen.) Production at a scale large enough to replace

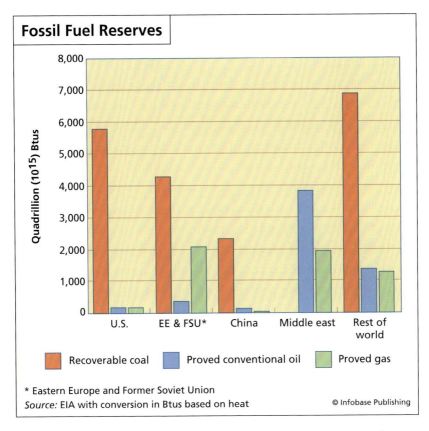

Fossil Fuel Reserves

While any fossil fuel can, in theory, serve as a feedstock for hydrogen production, only coal exists in sufficient quantities to power a hydrogen-based economy.

the petroleum used in the transportation sector would require an enormous amount of agricultural land. A committee appointed by the U.S. National Academies' National Research Council estimated that with current biomass technologies 650,000 square miles (1.7 million square kilometers) of farmland would be required to grow enough feedstock to produce sufficient hydrogen to replace the oil used in the United States in the transportation sector. At present there are only about 700,000 square miles (1.8 million km^2) of cropland in the United States. Even assuming substantial increases in agricultural productivity and the technology used to convert plant matter to hydrogen, the committee estimated that it would still require 280,000 square miles (730,000 km^2) of agricultural land to power a hydrogen economy.

Depending on biomass as the primary feedstock for a hydrogen economy would have profound impacts on the cost and availability of food as well as fuel, and (again) CO_2 would be produced as a by-product, although in the case of biomass, some of the CO_2 produced in the production and consumption of hydrogen would be removed from the atmosphere by the feedstock as it is grown. In the absence of sequestration, the CO_2 would return to the atmosphere when the biomass is consumed, creating a kind of carbon cycle. But if biomass were coupled with sequestration, the use of biomass as a feedstock would result in the net removal of CO_2 from the atmosphere, although the resulting hydrogen would probably be too expensive to serve as an oil substitute.

As noted previously, water can serve as a source of hydrogen, and electrolysis can be a method by which the hydrogen is produced. Despite the inefficiencies previously described, this method still draws interest because it is familiar, simple, and depending on the choice of primary energy source, produces no greenhouse gases. Photovoltaics, the technology that converts sunlight directly into electricity, is sometimes described as a good source of energy for hydrogen production via electrolysis because the primary energy source, the Sun, is free. But although the Sun is free, photovoltaic technology is extremely expensive. Moreover, a solar facility would be idle every night and operate at diminished efficiency every cloudy day. These kinds of reduced efficiencies drive up costs even more. In fact, photovoltaic generation is currently the most expensive method of producing hydrogen, so while one could, in theory, produce sufficient hydrogen using enormous arrays of photovoltaic cells operating no more than 12 of every 24 hours, the resulting hydrogen would be too expensive to serve as a transportation fuel.

Wind turbines can also be used to power the electrolysis process, but as with photovoltaic cells, staggering numbers of wind turbines would be needed, and as with photovoltaics, wind turbines are intermittent power producers. It would, therefore, be necessary to build many wind turbines in many different places so that when electricity production fails in one location due to lack of wind, there is a good chance that production can continue elsewhere. These sorts of redundancies make for a very expensive system. In fact, from the point of view of the wind turbine owner, the easiest way to maximize one's profits would be to sell the electricity from the turbines directly to the grid without converting the electricity into hydrogen. Consequently, in order to make the production of hydrogen economically attractive for the turbine owner, the hydrogen would have to be priced to take into account hydrogen production, storage, and transport as well as the prevailing price of electricity. Such a pricing policy would, again, result in very expensive hydrogen. (The same argument holds for using solar power.)

There are also thermochemical processes that can produce hydrogen gas through the use of high-temperature chemical reactions. The method receiving the most attention involves using water that is very hot—between 1,300°F and 1,800°F (700°C–1,000°C). Such high temperatures require very powerful heat sources. In order to produce the enormous amounts of hydrogen required for a hydrogen economy via thermochemical processes, proportionately enormous amounts of thermal energy would also be required. This makes nuclear energy the heat source of choice. The technology to build high-temperature nuclear reactors exists. A few have already been built, but none are designed for the production of hydrogen. (The current generation of commercial nuclear reactors in use in North America, where three designs—boiling water reactors, pressurized water reactors, and CANDU reactors, also known as pressurized heavy water reactors—comprise the entire market, are unsuitable for hydrogen production. They do not operate at high enough temperatures.) High-temperature nuclear reactors dedicated to the production of hydrogen would produce zero CO_2 emissions while simultaneously producing enormous quantities of hydrogen. But these machines would be very expensive, and the current demand for hydrogen is limited. Until a large hydrogen market has been established, there will be no large-scale production of hydrogen using nuclear reactors. The enormous costs of the reactors would be out of all proportion to the tiny potential profits. (Strategies for transitioning to a hydrogen economy are discussed in chapter 8.)

Finally, it is, at least in theory, possible to produce large quantities of hydrogen using microorganisms. This research, which is still in its very early stages, shows promise but is also many years away from commercial deployment.

Evidently, more research is needed into methods of hydrogen production.

But hydrogen production is only the first problem; the next barrier is storage.

ONBOARD HYDROGEN STORAGE

Assume that a method has been found to economically produce large amounts of hydrogen and assume an infrastructure has been built to distribute the hydrogen to many filling stations. The next problem involves storing enough hydrogen on board a car or truck to enable that vehicle to have a long enough cruising range to be useful. The most fundamental difficulties associated with onboard storage arise from two physical properties of the element hydrogen.

The first physical characteristic that distinguishes hydrogen from other potential fuels is its comparatively low density. The density of any material is the ratio between the mass of a representative sample of the material and the volume occupied by that sample. In most fields of science and engineering the units in which density is expressed are kilograms per cubic meter, but among researchers in the field of energy, it is common to see density expressed in

Honda filling station FCX. Most major automakers now have a hydrogen-powered car research program. *(National Hydrogen Association)*

pounds per cubic foot, kilograms per cubic meter, or occasionally kilograms per cubic foot. To facilitate comparisons of the densities of various materials, especially gases, the conditions—that is, the temperature and pressure under which the density measurements are performed—are standardized. Among those who do business or research in the field of energy, standard conditions are often defined to be 60°F (15.6°C) and 1 atmosphere pressure, and these will be the conditions described as "standard" throughout this chapter unless otherwise noted. (In many fields of science and engineering, standard conditions are expressed in degrees kelvin and bars pressure, and they are defined to be 273.15°K [that is, 0°C or 32°F] and 1 bar, which is equivalent to 0.9869 atmospheres.)

At standard conditions hydrogen is a gas, and one kilogram (2.2 pounds) of hydrogen occupies about 11.8 cubic meters (416 cubic feet). That is about eight times the volume occupied by one kilogram of methane, the principal component of natural gas, under

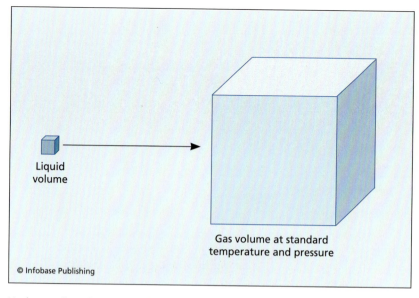

Liquid
volume

Gas volume at standard
temperature and pressure

© Infobase Publishing

Hydrogen liquid-to-gas-volume expansion ratio

standard conditions, and it is about 9,000 times as much volume as is occupied by one kilogram of (liquid) gasoline at standard conditions. It might seem, therefore, that in order to carry sufficient hydrogen gas to travel a meaningful distance, an automobile powered by hydrogen gas would, of necessity, have a very large tank. Such a large tank would either require the car to be very large or require that the tank occupy the entire interior of the car. Neither alternative is very attractive to the consumer.

Another possibility for storing hydrogen on board a car involves compressing the hydrogen gas until it is at a very high pressure. But even at 330 times atmospheric pressure hydrogen remains a gas, and its density, even at this high pressure, is, when compared to other energy sources, still low. Comparing two tanks with similar volumes, one carrying gasoline and the other carrying hydrogen gas compressed to 330 times atmospheric pressure, the hydrogen tank holds only about one-tenth of the energy that the gasoline tank holds. As a consequence a car that runs on hydrogen gas must have a large fuel tank *and* the hydrogen must be stored at very high pressures in order to have the same cruising range as a similar gasoline-powered car. It is easy to see that there are potential safety issues involved with driving at highway speeds in a car with a fuel tank containing highly flammable gas compressed to hundreds of times atmospheric pressure. There are also safety issues associated with storing high pressure hydrogen gas at service stations and with allowing untrained consumers to fuel their cars with hydrogen at the high pressures required for adequate storage.

Another alternative to the problem of storing adequate amounts of hydrogen fuel in an automobile is to use liquid hydrogen instead of hydrogen gas. Liquid hydrogen is much denser than hydrogen gas; a tank of liquid hydrogen would occupy a much smaller volume than one holding the same mass of gas even when the gas is compressed to several hundred atmospheres. This approach is already used for other types of gaseous fuels. Propane, for example, is a common home heating fuel that is also used in outdoor gas grills, and it provides a useful standard for comparison.

Trucks carry propane from house to house for use as a home heating fuel. At ordinary temperatures and at atmospheric pressure propane is a gas. In order for each truck to carry as much propane as possible, the propane is liquefied to increase its density. In the case of propane, however, the liquefaction process is easy. At atmospheric pressure propane's boiling point is –43.8°F (–42.1°C). To liquefy propane, one can reduce its temperature until the temperature of the gas is below the boiling point. The gas condenses, and the resulting liquid is stored in high-pressure containers. The liquid is then allowed to warm until it is at the temperature of the surrounding environment. Because the storage containers are sealed, the gas is unable to expand and so remains in liquid form but at an elevated pressure. In theory, the same process works for hydrogen, but hydrogen's boiling point is –422.8°F (–252.8°C). It is, therefore, much more expensive to liquefy hydrogen because much more energy is required to lower its temperature by the hundreds of degrees required. Currently, in order to liquefy a given mass of hydrogen, about 30–40 percent of the energy content of the hydrogen is required for the liquefaction process. In other words, much of the energy in the hydrogen fuel, the energy that one might hope to apply toward transportation, would instead be used to reduce its temperature below its boiling point. It is worth noting that even when hydrogen is in liquid form, gasoline still has four times the

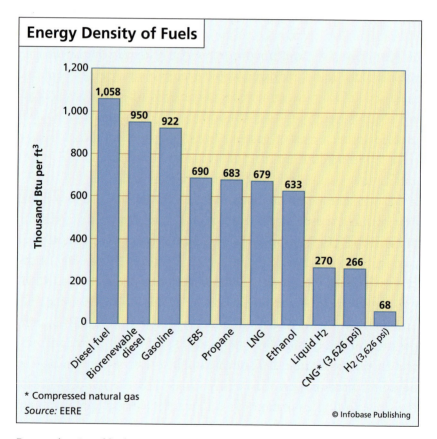

Energy density of fuels. Compared to more conventional fuels, a tank of hydrogen gas does not contain much energy. *(EERE)*

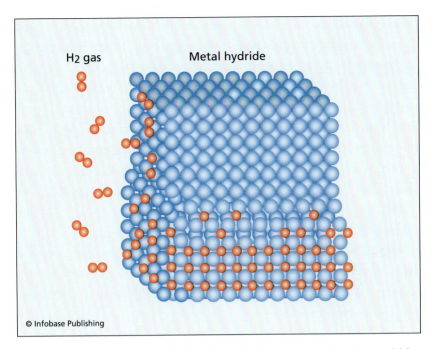

The idea behind a metal hydride is that the hydrogen molecules would be stored within the molecular lattice like water within the pores of a sponge.

energy content on a per-volume basis because even when liquefied, hydrogen is not especially dense.

Other storage technologies have been proposed. One possibility is to store hydrogen as a liquid hydrocarbon—gasoline, for example—and to use an onboard reforming process, similar in concept to the process by which methane is reformed, to produce the necessary hydrogen while the car is being driven. Hydrocarbon fuels contain substantial amounts of hydrogen. They are relatively easy to transport and store, and, of course, the infrastructure needed to transport and store gasoline is already in place. Including a device on board the car to reform a liquid hydrocarbon fuel would, however, substantially increase the price of the car, its complexity, and, presumably, the cost of maintenance. It is difficult to see how cars manufactured with onboard *reformers* would be competitive with gasoline-powered cars at any time in the near future. Moreover, there is currently no practical way to store for later sequestration the carbon that an onboard reformer would produce. Cars with onboard reformers would, therefore, continue to emit greenhouse gases much as today's gasoline-powered cars do.

In theory, hydrogen could also be stored in specialty materials that retain hydrogen atoms in a way that is somewhat analogous to the way that a sponge retains water. The hydrogen could be withdrawn from the storage medium for use as needed. Many such materials are known, but they tend to be expensive, and their storage capacity is currently insufficient to enable the consumer to use a hydrogen powered vehicle in the same way that gasoline powered vehicles are currently used. At present there are no easy answers to the problem of storing hydrogen for use as a transportation fuel. More research is needed.

ENERGY CONVERSION

The way that hydrogen is used as a fuel is just as important as its physical properties. Hydrogen is versatile. It can, for example, be combusted in a way that is similar to that of natural gas, and light vehicles—cars and trucks—with internal combustion engines have been modified to burn hydrogen just as some cars and trucks have already been converted to use natural gas. But few see burning hydrogen in an internal combustion engine as a viable option because it is inefficient. An engine that burns hydrogen will convert much less than 50 percent of the thermal energy produced during combustion into mechanical energy. Given the inefficiencies already involved in producing and storing hydrogen, this is probably one inefficiency too many.

Currently, the most efficient way of converting the chemical energy of hydrogen into mechanical energy is to first convert the hydrogen into electrical energy via a device called a fuel cell. The electricity can then be used to drive an electric motor. Some fuel

Gemini 5 was the first U.S. spacecraft to attempt to use a fuel cell. *(NASA)*

cells are two or even three times more efficient in converting chemical energy into mechanical energy than are heat engines.

The key to the fuel cell's efficiency lies in *direct* conversion of chemical energy into electrical energy—that is, there is no working fluid, no piston, and no turbine. Although some fuel cells produce large amounts of thermal energy, they are not heat engines. They have, in fact, more in common with batteries than heat engines, and as with batteries, the fuel cell is not a new idea. The British physicist and jurist Sir William Robert Grove (1811–96) developed the first fuel cell, and throughout the next century, fuel cell research continued sporadically. Efforts to improve the fuel cell intensified beginning in the 1960s during the early days of the space race between the United States and the Soviet Union. NASA, in fact, first tested fuel cells in space in 1965 during the *Gemini 5* mission. (The Gemini Project was a series of space missions, each of which consisted of two astronauts circling Earth in a small space capsule. The purpose of each Gemini mission was to field-test the technology and techniques needed for successful space exploration, especially for the upcoming Apollo missions, which depended on fuel cells for power.) In part because of their efficiency, fuel cells are ideal for space exploration. They are better than batteries because they weigh less than batteries with the same power rating. Unlike batteries, they do not run down. They can provide a steady supply of electricity for as long they receive a steady supply of hydrogen and oxygen. As an additional advantage, the water that is produced as a by-product of the fuel cell process is clean enough to drink. In fact, early astronauts drank the water produced by their onboard fuel cells, and they still do. But earthbound applications for fuel cells, either as stationary sources of electric power or as sources of power for cars and trucks, have proved far more elusive.

Many different types of fuel cells now exist, and they can be classified in many different ways. Performance characteristics vary from type to type. Some fuel cells, for example, operate at 1,800°F (1,000°C), and others operate at 180°F (80°C). The electrical output of each type of fuel cell together with its durability determine the applications for which it is best suited. The fuel cell of most interest in this volume is called a polymer electrolyte membrane fuel cell—also known as a proton exchange membrane fuel cell and more frequently called a PEM fuel cell. The PEM fuel cell currently shows the most promise for automotive applications. Many of the same concepts employed in the design of the PEM fuel cell are, however, used in other types of fuel cells.

The heart of the PEM fuel cell consists of three pieces: the anode, the cathode, and a piece of plastic sandwiched between the anode and cathode called the electrolyte, which is a membrane that is as thick as a few sheets of paper. When the membrane is damp, protons, which are positively charged particles usually found in the nuclei of atoms, can migrate through it—hence the name "proton

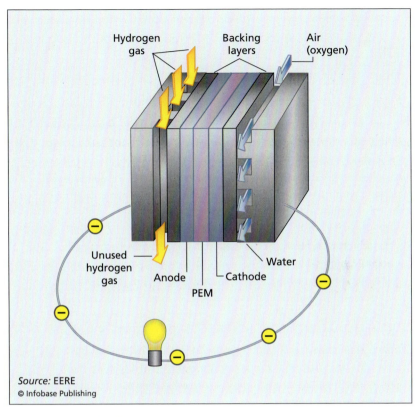

Anatomy of a fuel cell *(EERE)*

exchange membrane"—but the PEM has the additional property that electrons are unable to move through the membrane. When damp, the membrane functions as a sort of gate between the anode and the cathode, allowing protons, but not electrons, to pass from one side to the other.

The protons that migrate through the damp membrane come from the hydrogen gas that is furnished to the anode. Recall that a hydrogen atom consists of a single proton and a single electron. As hydrogen is supplied to the anode, hydrogen atoms dissociate into

(continues on page 140)

Hydrogen Safety

Demonstrating that hydrogen is safe will not, of itself, guarantee that hydrogen-fueled technology will win widespread public acceptance. But if it can be shown that hydrogen is an unsafe fuel, the public will quickly lose interest in it as an alternative to gasoline. The safety issues associated with hydrogen are more complex than those associated with other, more common fuels. Consider the following issues:

1. Compared with other combustible gases, hydrogen burns under an unusually wide range of concentrations. Hydrogen mixed with air will burn when it occupies anywhere from 4 percent to 75 percent of the volume in question. In practice, this means that a hydrogen leak is dangerous under a broader range of conditions than those associated with natural gas, propane, or gasoline vapors (see the accompanying diagram). On the positive side, hydrogen gas tends to diffuse more rapidly in air than other combustible gases, and leaks, provided they occur in open spaces, soon disperse.

2. Hydrogen ignites easily—far more easily than other gaseous fuels. A very weak spark, a spark that would not ignite other common gaseous fuels, is enough to start a hydrogen flame. From a practical point of view, the ease of igniting hydrogen together with its wide limits of flammability mean that refueling a hydrogen vehicle poses special safety challenges, as does leaving a hydrogen-powered vehicle in an enclosed garage.

3. A flame moving through a hydrogen-air mixture travels much more rapidly than a flame moving through a natural gas–air mixture or a propane-air mixture. Therefore, in a confined volume, a hydrogen ignition is much more likely to become a hydrogen explosion than it is for any of the other common gaseous fuels.

4. Hydrogen has no smell. Neither does natural gas, of course, but a chemical called an odorant is added to natural gas to warn the user in the event of a leak. Most odorants do not mix well with hy-

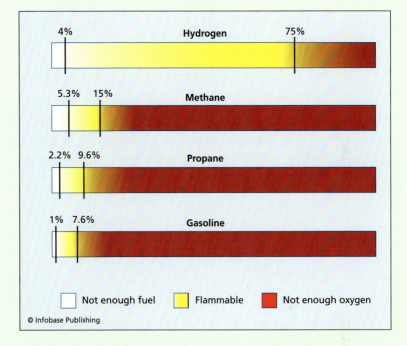

Hydrogen burns under a much wider set of conditions than other, more conventional fuels.

drogen, and the common ones would ruin any fuel cell in which they were used. Fuel cells require a very pure hydrogen source in order not to foul the catalyst.

5. A hydrogen flame is a very pale, almost invisible blue. In theory, it is possible to add a chemical to make the hydrogen flame more visible, but in practice these chemicals would also foul the catalyst in a fuel cell. Fuel cells produce no flames, of course, but

(continues)

(continued)

adding a chemical to hydrogen to make the flame more visible would also mean that hydrogen distributors might have to offer separate hydrogen supplies for fuel cells and for applications that involve burning hydrogen.

These facts do not mean that hydrogen cannot be handled safely. Indeed, hydrogen is handled safely every day—by specially trained individuals. The safety procedures used to handle gasoline, natural gas, and propane will not, however, be adequate for consumers in a hydrogen economy. Many common practices—storing automobiles in unventilated garages, for example, and allowing consumers to fuel their own vehicles—may have to change.

(continued from page 170)

their constituent protons and electrons. In symbols the reaction that occurs at the anode is written in this way:

$$2H_2 \rightarrow 4H^+ + 4e^-$$

The symbol H^+ denotes the nucleus of a hydrogen atom, which consists of a single (positively charged) proton, and the symbol e^- represents a single (negatively charged) electron. To describe this another way: Each two hydrogen molecules disassociate into four protons and four electrons.

As the protons are produced at the anode, they begin to migrate through the electrolyte to the cathode. Because the electrons are unable to follow the protons through the membrane, they move through an external circuit that connects the anode to the cathode. It is the movement of electrons through the external circuit that is the reason that the fuel cell was constructed. This flow of electrons is electricity.

At the cathode, on the other side of the fuel cell, oxygen is provided in the form of oxygen molecules or O_2. These oxygen molecules combine with the protons that have migrated through the electrolyte and the electrons that have migrated through the external circuit to produce water, or H_2O. In symbols, the reaction at the cathode is expressed in this way:

$$O_2 + 4H^+ + 4e^- \rightarrow 2H_2O$$

The hydrogen atoms are reconstituted and become bound to the oxygen atoms to make water molecules.

Unfortunately, these reactions occur at such slow rates within the fuel cell that the cell, as described, is a poor source of electrical power. Not enough current is produced per unit time. To increase the amount of electrical current produced per unit time, the fuel cell is made with a *catalyst,* a substance that increases reaction rates. Currently, the most efficient catalyst for this purpose is the very expensive metal platinum.

Together, the electrolyte, the cathode, the anode, and the catalyst are called the electrode/electrolyte assembly. The electrode/electrolyte assembly is itself sandwiched between backing layers that control the manner in which the hydrogen and oxygen are supplied to the anode and cathode respectively. It is important that these gases be supplied in a carefully controlled way for maximum efficiency, and this is the function of the backing layers, which ensure that the entire surface of the anode is supplied with hydrogen and the entire surface of the cathode is supplied with oxygen. Finally, the backing layers are themselves bracketed by current collectors that channel the hydrogen and oxygen to the backing layers and also provide a conductive surface through which the electricity can flow. The electrode/electrolyte assembly, the backing layers, and the charge collectors form a single, not especially powerful, fuel cell.

To increase the power rating of the fuel cell, one adds more cells to form a fuel cell stack—the more cells in the stack, the more powerful the stack is.

The design of a PEM fuel cell may seem simple, but to make it work efficiently at a cost the consumer can afford has proven to be extremely difficult. One can find many articles written during the 1970s and 1980s in which the writer confidently predicted that homes and cars would all be powered by fuel cells by the year 2000, but the time when the fuel cell replaces the internal combustion engine seems no closer now than it did when those articles were written.

Despite their difficulties, fuel cells remain an extremely promising technology. Notice that, as with all fuel cells, PEM fuel cells have no moving parts; they are quiet, and provided they are supplied with hydrogen and oxygen free of impurities, they can operate for prolonged periods of time without maintenance. They are efficient in the sense that they convert a large fraction of the chemical energy contained in the hydrogen into electricity. In an automobile, the electricity from fuel cell stacks can be used to drive an electric motor. The result is, at least in theory, a pollution-free vehicle.

Larger stationary power plants can also be constructed using fuel cells. Fuel cells are, in theory, attractive to utilities because they enable utilities to meet the ever-growing demand for power with minimal financial risk. The reason is that fuel cells are modular. They can be combined, or stacked, in the same way that batteries are combined to produce larger amounts of power. One need not build a large power plant in the hope that there will be enough demand for power to justify its expense. Instead, one can periodically add stacks of fuel cells and allow supply to grow in step with demand. And as previously mentioned, fuel cells supplied with hydrogen and oxygen produce only water as a by-product. At first glance, fuel cells seem to be the ideal solution to the problems of power generation. Why, then, are they not more widely used? What would it take to move the world's economy away from one based on fossil fuels to one based on hydrogen?

Some of the disadvantages that inhibit the widespread commercial acceptance of fuel cell technology are not apparent in the NASA applications described earlier. Spacecraft technology is a narrow, highly specialized branch of engineering, and with respect to NASA, price, while important, is seldom the deciding factor in making purchasing decisions. Moreover, once installed in a spacecraft, expert technicians adhere to exacting maintenance schedules to insure that the equipment will operate in accordance with design specifications. Consumer products, by contrast, are subject to a very different set of constraints. Maintenance schedules are often ignored by consumers, who, nevertheless, expect the products that they have purchased will continue to operate efficiently. And technicians responsible for maintenance and repair of consumer devices have widely varying skill levels. The equipment and technical skill needed to maintain fuel cells may not be available in the region in which the consumer lives, and unlike NASA, cost considerations are often *the* decisive factor for consumers when making purchasing decisions.

Despite the drawbacks of fuel cells as consumer items, a number of companies around the world, small and large, have begun to introduce them into the marketplace. Costs for these units are sometimes heavily subsidized by governments. Both the U.S. departments of Energy and Defense, for example, have funded fuel cell demonstration projects. As a consequence of this support, fuel cell research and development continues. Some believe that it is just a matter of time before fuel cells are as commonplace as internal combustion engines.

The Hydrogen Economy

The phrase "the hydrogen economy" seems first to have been used during the 1970s by engineers at General Motors (GM). Inspired by breakthroughs in fuel cell technology at NASA and under pressure to respond to the embargo imposed by the Organization of Petroleum Exporting Countries (OPEC) in 1973, engineers at GM began to consider the possibility of replacing gasoline with hydrogen. Today, decades later, automotive engineers at many research laboratories around the world are still considering it. The reasons are clear. Oil prices have been volatile for decades, and supplies are growing ever tighter as China and India, which together are home to one third of the world's population, consume ever-larger quantities of oil. Oil consumption is also increasing, albeit more slowly, in many developed countries. (The same statements are true for natural gas. And the consumption of coal is also increasing, although in contrast to natural gas and oil, coal is plentiful and relatively

Chevrolet Sequel hydrogen car. Research into hydrogen-powered cars began in the 1970s. *(General Motors)*

inexpensive.) It is, therefore, difficult to see how the rate at which carbon dioxide is added to the atmosphere can decrease over the next few decades. In fact, as an ever-larger proportion of the world's population becomes prosperous, it is not at all clear that even stringent conservation measures would be sufficient to halt the rate of increase of fossil fuel use.

A hydrogen economy holds out the hope that it is possible to maintain and even increase the rate at which economies grow and simultaneously diminish the environmental effects associated with that growth by simply changing fuels. But how could the nations of the world make the transition from an economy based on fossil fuels to an economy based on hydrogen? What needs to happen, and in what order? These are the questions addressed in this chapter.

THE TRANSITION TO A HYDROGEN ECONOMY

Moving from gasoline to hydrogen is a difficult and expensive proposition. As described in chapter 7, there are a number of technical hurdles that must be overcome before hydrogen becomes a practical fuel.

Hydrogen is still expensive to produce, difficult to store in reasonably sized containers, and fuel cells, theoretically the most efficient devices for using hydrogen fuel, remain too expensive and too unreliable for most practical applications. But even if all these technical hurdles could be overcome, there remains an additional problem that was only briefly mentioned in the preceding chapter. Because there is currently very little demand for hydrogen, there is no reason to create a hydrogen supply infrastructure, which would be very expensive; and because there is no hydrogen supply infrastructure, there is no reason to purchase the often-expensive devices that use hydrogen as a fuel. How, then, can a hydrogen economy be created?

Some writers compare the possible transition from oil to hydrogen with the transition that occurred a century ago when coal, used to power trains and ships, was thrown over for gasoline. But this comparison is of limited value because the two transitions are quite different. One hundred years ago, even before there was much demand for gasoline, gasoline supplies were already plentiful. Refineries produced it almost as a by-product of the process by which they produced kerosene, which, at the time, was the petroleum product for which demand was highest. Therefore, prior to the widespread adoption of the automobile, gasoline was already cheap and easily available. By contrast, hydrogen supplies are currently tight and comparatively expensive.

A second difference between hydrogen and gasoline lies in functionality. Early cars were functionally different from other forms of transportation then available. Early automobiles could, for example, carry passengers to places that trains could not reach. Car owners could travel on schedules that they otherwise could not follow. Broadly speaking, gasoline-powered vehicles were forms of personal transport; coal-powered vehicles were forms of mass transport. Gasoline made a new type of mechanical transport possible.

Today, engineers hoping to transition from gasoline to hydrogen want to retain functionality rather than change it. Their goal is to build hydrogen-powered cars with characteristics similar to those of gasoline-powered cars. They want hydrogen-powered cars to be as safe and as easy to refuel as gasoline-powered cars, and they want hydrogen-powered automobiles to have cruising ranges roughly equal (or better) than those of similarly priced gasoline-powered automobiles. They want hydrogen-powered cars to operate *as if they were* gasoline-powered cars. This is a harder goal to accomplish because gasoline-based automotive technology is very highly developed. The transportation niche for which hydrogen power engineers are designing is already occupied.

There are still other barriers to the introduction of hydrogen as an important transportation fuel. During the last century, a trillion-dollar infrastructure was created to support gasoline-powered vehicles. There is currently not a comparable system for supporting hydrogen-powered vehicles, and given the expense involved in building one, it is unlikely that such a system will be in place soon. Also, gasoline-based technology is continually being improved. A hydrogen-powered vehicle that is competitive with a conventional gasoline-powered vehicle may not be competitive with a more fuel-efficient gasoline-electric hybrid or with so-called plug-in vehicles, battery-powered vehicles with gasoline backup. Gasoline-based technology presents a moving target for those wishing to introduce hydrogen-powered vehicles. *Even under the most favorable conditions,* therefore, hydrogen powered vehicles may not supplant gasoline-powered automobiles for several decades—and perhaps much longer.

There is also the problem of short-term profits. In a market economy, companies will not commit the enormous reserves of capital and expertise needed to build a large-scale hydrogen infrastructure unless there is a reasonable assurance of quick profits, and few consumers would be willing to purchase hydrogen-powered vehicles unless they feel confident that fuel would be available when and where they needed it. Under these circumstances, what can be done to facilitate movement toward a hydrogen economy?

One possibility is to build a so-called distributed system that would manufacture hydrogen on-site at individual filling stations. This system would bypass the need for large centralized hydrogen production facilities and an extensive and expensive infrastructure for the transport of hydrogen. Even under the best of circumstances, a distributed production system could not produce large amounts of hydrogen, but it could fuel a small number of vehicles. It would be a relatively inexpensive and low-risk way to start the transition. The disadvantages of a distributed production system are that hydrogen produced on-site would be significantly more expensive than hydrogen produced at a centralized production facility because distributed production cannot make use of economies of scale.

In *The Hydrogen Economy: Opportunities, Costs, Barriers, and R&D Needs,* a study written for the U.S. National Academies National Research Council, the authors envision a distributed hydrogen production system for light vehicles that would work as follows: Each small-scale production facility would produce about 950 pounds (430 kg) of hydrogen per day, enough to supply the needs of 800 automobiles. At this scale the committee considered two possible feedstocks, natural gas and water, because these two fuels are the most readily available for small-scale production.

In the case of natural gas, hydrogen would be produced via the process of steam reforming. Natural gas is an attractive option because an infrastructure for supplying natural gas to filling stations already exists in many areas, and steam reforming of natural gas is a well-understood technology. With respect to the use of water as a feedstock, the committee considered three methods of producing hydrogen, all of which depend upon electrolysis. The three different methods of producing hydrogen by electrolysis are distinguished by the technology used to supply the electricity. The methods were photovoltaics, wind, and (more broadly) electricity taken directly from the electrical grid no matter how it was produced. Using electricity from the grid would be fairly expensive. The other sources of electrical power considered by the committee make direct use of renewable energy. One would think that the "free" energy one obtains by converting sunlight and wind into electricity would be inexpensive, but as mentioned in chapter 7, the reverse is true. Despite decades of research, photovoltaic technology is still fairly expensive, and the equipment would be idle more often than not because the array would produce no electricity at night. Consequently, one must either build a very large array to ensure that sufficient hydrogen will be produced during favorable periods, or one must purchase electricity from the grid to cover any shortfalls. The situation is further complicated by the fact that filling stations are rarely sited in the sunniest locations. Therefore, any hydrogen produced by a distributed system relying on photovoltaics would be extremely expensive. Essentially the same arguments apply to wind power.

What about the effects of distributed generation schemes on the environment? With respect to small-scale steam reforming, the sequestration of CO_2 produced by the reforming process is impractical, and so all CO_2 would be vented to the atmosphere. The use of grid-based electricity would also generate substantial CO_2 emissions if the electricity were provided by a generating station that used coal or natural gas and did not sequester its CO_2 emissions. (As of 2009, no fossil fuel power plants in the United States sequester their CO_2 emissions.) The system would, however, be emissions-free in areas served by nuclear power plants.

The committee also tried to estimate the costs involved. Their analysis indicates that distributed generation of hydrogen using natural gas is by far the most cost-efficient of the four small-scale hydrogen production methods, but even with steam reforming of natural gas, the fuel costs for driving a hydrogen-powered car would be about twice as much per mile as one would spend driving a gasoline-electric hybrid vehicle. Using hydrogen produced from photovoltaics, the fuel costs incurred by the consumer who drove a hydrogen-powered vehicle instead of a gasoline-electric hybrid would be about five times as great. These are only estimates, and they were based on assumptions about the future price of oil, the price of natural gas, the price of photovoltaics, and the ultimate efficiency of vehicles that have yet to be built, so there remains a good deal of uncertainty about details. Indeed, the price of oil has been very volatile in recent years, a situation aggravated in the United States by large fluctuations in the value of the dollar relative to other currencies. But the trend is still clear: The cost of hydrogen would, even now, have to be heavily subsidized, at least initially, before many consumers would, or even could, move toward hydrogen-fueled cars. (The situation is further clouded by the effects of increasing demand for hydrogen. If there were enough hydrogen filling stations using natural gas as a feedstock, one would expect that the cost of natural gas would increase in response to increasing demand. The final cost of the resulting hydrogen would reflect these additional costs.)

Assuming that a distributed system was profitable enough to warrant expansion, a possible next step in transitioning to a hydrogen economy would be the construction of midsized production facilities. In this case the authors of *The Hydrogen Economy* considered biomass as the feedstock. Biomass is a term that includes agricultural residues as well as specially grown crops. The biomass could be converted into a gas, which would be reformed to produce a stream of hydrogen in a way that is conceptually similar to what is done with natural gas. The committee believed that midsized plants using biomass as a feedstock would have a maximum production capacity of 53,000 pounds (24,000 kg) of hydrogen per day. Operating at a more realistic capacity of 90 percent of their maximum rate of production, these plants would produce about 47,600 pounds (21,600 kg) per day, enough to fuel about 40,000 cars. Each such plant would produce about 50 times as much hydrogen as the small plants in a distributed system—a big increase, but not enough to warrant building a pipeline infrastructure such as the one that already exists for natural gas. Consequently, the hydrogen would be transported to the point of sale in liquid form using tanker trucks or trains, a system that would, for the reasons described in chapter 7, be fairly expensive.

The resulting price of hydrogen would, in fact, be the main disadvantage to using biomass as a feedstock in mid-scale facilities. The price of hydrogen produced at midsized plants from biomass would probably be comparable to the price of hydrogen produced at the much smaller distributed production plants, in part, because the amount of hydrogen obtainable from gas produced from biomass is smaller than the amount of hydrogen obtainable from an equal amount of natural gas when measured on a per-mass basis. A great deal of energy would have to be expended to liquefy the hydrogen to prepare it for transport, and the costs associated with the production and collection of biomass feedstock are also high.

Although the economic justification for the use of biomass as a hydrogen feedstock is not compelling, there are some good environmental reasons for using biomass. When biomass is used as a hydrogen feedstock, some of the CO_2 released into the atmosphere during hydrogen production is removed from the atmosphere by growing the feedstock. Biomass-based hydrogen can, therefore, result in reduced net carbon dioxide emissions. But facilities that produce hydrogen from biomass have the potential for doing much better than "reduced net carbon dioxide emissions." Because the process of creating the hydrogen is done in a centralized and controlled way, it is possible, at least in theory, to take the carbon dioxide generated during the production of the hydrogen and sequester it. In this way, the production of hydrogen from biomass at a midsized facility actually has the potential to permanently remove CO_2 from the atmosphere: The plants remove CO_2 from the atmosphere as they grow, and that CO_2 is permanently sequestered during the hydrogen production process.

By beginning with small distributed facilities and moving to the much larger midsized facilities as consumer demand increases,

(continues on page 154)

How Long until the Hydrogen Economy?

The first thing to keep in mind is that there are very few people who want a hydrogen economy as such. With respect to transportation, the hydrogen-based application that attracts the most attention, what people want is a technology that is nonpolluting, inexpensive, convenient, and safe to use. That may turn out to be cars driven by hydrogen-powered fuel cells or it may mean hydrogen-powered internal combustion engines, but it may turn out to be another technology entirely—battery-driven cars, for example. A revolution in battery technology would quickly short-circuit the hydrogen economy.

Assuming that no other technology "short-circuits" the hydrogen economy, there are still a number of difficult technical problems that must be overcome if the hydrogen economy is to become a reality. These are described in the main body of the text, and include the following:

- producing hydrogen at an affordable cost
- producing commercially significant amounts of hydrogen without devastating the environment
- finding a way to store sufficient amounts of hydrogen onboard so that the hydrogen-powered vehicle will have a cruising range that is roughly comparable to that of a conventional vehicle
- if hydrogen is carried onboard in the form of a gasoline-like fuel, finding a way to use the fuel in a way that causes no more pollution than a conventional gasoline-powered engine
- the creation of a durable and affordable fuel cell for use in automobiles and trucks

All of these problems have proved resistant to solution.

Assuming that all of them are solved within the next decade or two, there would still be no reason for the consumer to buy a hydrogen-powered car, because initially, there would be so few places to buy fuel. There is no reason for a businessperson to build a fueling station without a market, and just one fueling station would have little value anyway. In order for a hydrogen-powered car to be a useful form of transportation, there must be many fueling stations. The construction of such an infrastructure would take many billions of dollars and many years to accomplish.

Meanwhile, there would still be a trillion-dollar petroleum infrastructure in place. Faced with competition from hydrogen- or battery-powered cars, those with money invested in petroleum would do their best to compete. Competition might consist of building automobiles with characteristics designed to appeal to those thinking about buying an alternative-fuel vehicle, or it might consist of directly impeding companies engaged in building alternative-fuel vehicles. Historically, both strategies have been employed by entrenched business faced with new competition. Oil-dependent businesses will have the means to pursue either strategy.

To gain some perspective, consider the demise of the coal gas industry. As described in chapter 1, there was a time when coal gas was used almost exclusively for lighting, but Edison's invention of the electric light did not spell the end of coal gas. The reliability of a mature (coal gas) technology, innovations in gas lighting, and the willingness of coal gas manufacturers to market their fuel for new applications, especially for heating and cooking, enabled an industry that was a heavy polluter and, in retrospect, demonstrably inferior to both electricity and natural gas, to continue to prosper for a further half-century after the invention of the incandescent light. It was not until 50 years after the invention of the electric light that the coal gas industry entered into a period of steep and permanent decline.

Technical difficulties, economic difficulties, and historical precedent all indicate that it is not unreasonable to expect that the hydrogen economy, if it ever comes to pass, will not occur for at least a generation.

(continued from page 151)
the risks assumed by investors are minimized and investors and consumers alike can ease their way into a true hydrogen economy.

CHARACTERISTICS OF THE HYDROGEN ECONOMY

Suppose, now, that reasonable solutions are found for the technical problems that have so far prevented the switch from fossil fuels to hydrogen. In particular, assume the following:

> ➤ the technology needed to manufacture the requisite volumes of hydrogen at a reasonable cost is ready to be implemented;

> ➤ a way is found to manufacture sturdy, inexpensive fuel cells for use in the transportation sector and in the power generation sector; and

> ➤ economical solutions are found for the problems of safely transporting and storing hydrogen.

How expensive would it be to implement a hydrogen economy? In what ways would an economy based on hydrogen be different from one based on fossil fuels?

Consider, first, the impact on the world economy of a transition from oil to hydrogen. It is in the transportation sector that a switch to hydrogen would have the greatest impact. Whether measured by volume shipped or dollars exchanged, oil constitutes the biggest single item of international trade. Since 1973, when OPEC nations created a new economic world order by taking control of their own oil fields, there has been an enormous transfer of wealth from oil-importing nations to oil-exporting nations. That transfer of wealth continues today. If hydrogen were to become the principal transportation fuel, oil-exporting nations would experience an economic disaster because most oil is used in the transportation sector.

Today, oil refineries such as the one pictured here are some of the main consumers of hydrogen. *(Europa Valve Ltd.)*

Eliminate the demand for gasoline and diesel fuel, and there would be little left of the oil business, one of the world's largest businesses. This would have a profound impact on the economies of many nations, but oil-exporting nations, most of which have no other significant source of income, would face economic collapse.

The transition to a hydrogen economy would, however, probably happen slowly. As hydrogen became competitive with oil as a transportation fuel, both the demand for and the price of oil would decrease. As the price of oil diminished, its competitive position relative to hydrogen would improve, thereby reducing the demand for hydrogen. Whatever negative environmental and economic effects are associated with the world's extreme dependence on oil—and there have been many and they have been severe—oil remains attractive as a transportation fuel because gasoline, diesel fuel, and jet fuel are so easy to transport, store, and burn. The time required to effect the transition would, therefore, probably be measured in decades.

From an environmental point of view, a hydrogen economy is often described as environmentally benign because the use of hydrogen in fuel cells produces only water as a by-product, and burning hydrogen together with pure oxygen likewise produces only water as a by-product. Of course, practical combustion devices use air rather than pure oxygen, but even in practical devices, the amount of pollution produced by burning hydrogen is very small. In contrast to fossil fuels, most of the environmental problems associated with a hydrogen economy would arise not from the consumption of hydrogen but from its production.

Coal is often mentioned as the most likely primary energy source for powering a hydrogen economy. Many people are initially surprised by the prominent place that coal has in planning for a hydrogen economy because coal is often and justifiably described as dirty—dirty to produce and dirty to burn. But hydrogen gas must be manufactured before it can be used, and the production process requires an abundant feedstock and a large primary energy source. Coal can serve both functions.

Must the use of coal in a hydrogen economy result in widespread environmental damage and substantial health risks? This is less clear. Certainly, the process of mining coal involves significant risks to miners, and many miners have suffered permanent injury or even death in the practice of their profession. While safety standards have improved as a result of government regulation and advocacy by the United Mine Workers of America union, much remains to be done, and injury and death remain an integral part of the business of mining. There is also the threat of local environmental damage. Mining companies have a long history of questionable environmental practices. Perhaps the most controversial practice today is that of mountaintop removal, a practice that involves removing the top of a coal-bearing mountain and filling a nearby valley with the non–coal-bearing rock and soil, called overburden, that must be removed before the coal is exposed. Other problems include acid run-off from mines and the disruption of aquifers from underground mining activities. Some of this may be unavoidable. There is no easy way to extract the amount of coal required to fuel today's power plants, and there would be no easy or environmentally benign way to extract the tremendous volumes of coal required to power a hydrogen economy.

The burning of coal has long been associated with poor air quality. Sulfur dioxide and nitrogen oxides, both of which are produced in abundance when coal is burned, are important contributors to air pollution. Certain heavy metals, especially mercury, are released by burning coal and can now be found widely distributed throughout the environment because they were released by coal-burning power plants. Today, various antipollution devices capture most of the pollution produced by burning coal before it reaches the environment. The days of smokestacks billowing enormous black clouds are gone. But the installation of sophisticated antipollution equipment only means that a very high *percentage* of most pollutants are trapped before the products of combustion are vented to the atmosphere. So much coal is burned that the total amount of pollution reaching the atmosphere through the smokestack remains significant even though it is a small percentage of the pollution produced during combustion, and the cumulative effect of what does escape can still be large. For example, a great deal of coal is burned each day in the U.S. Midwest to produce electricity, and despite the antipollution equipment installed on those midwestern plants, a number of environmental problems in the Northeast section of the United States—particularly in the Adirondack region in upstate New York—can still be traced to pollutants emitted from power plants situated in the Midwest.

Perhaps the most important environmental consequence currently associated with a heavy reliance on coal involves the emission of large quantities of greenhouse gases, principally CO_2. The large-scale emission of CO_2 is certainly the only practice associated with coal consumption that has the potential to affect the entire planet. Across the world, all of the CO_2 produced during the combustion of coal is vented directly to the atmosphere. If current technology is used to produce hydrogen from coal, about 20 times as much carbon dioxide will be emitted during the process of hydrogen production as hydrogen is produced when the two gases are measured by mass. Because so much hydrogen would be required to power a hydrogen economy, the production of the accompanying CO_2 promises to be a tremendous environmental burden. A switch to a coal-based hydrogen economy would involve an enormous increase in the production of greenhouse gases. A hydrogen economy based on coal might better be called a carbon economy.

The solution to the problem of CO_2 emissions from the use of coal would probably involve sequestration technology. Because carbon sequestration is based on the idea that the carbon would be permanently injected into geological formations (or perhaps deep beneath the surface of the ocean), it would be necessary to use highly stable geological formations with the capacity to retain staggering amounts of CO_2 indefinitely. Sequestration on such a scale may be possible, but no nation has any real experience with an antipollution project of this magnitude. Oil companies have long injected CO_2 into the ground to increase oil recovery rates, but these projects are comparatively small and done with little concern for whether the CO_2 remains in the Earth indefinitely. With respect to sequestration as an environmental strategy, the leader in this field has been Norway's state-owned oil company, Statoil, which since 1996 has injected about a million tons of CO_2 per year in a geological formation 2,600 feet (800 m) beneath the North Sea floor. Their experience has shown that at least under the circumstances prevailing where Statoil is working, and at the relatively small scale at which they are working, sequestration of CO_2 can be done economically and in an environmentally sound way. Is coal the right feedstock and the right primary power source to supply a hydrogen economy? The only way to answer the question is to compare coal with the alternatives.

As previously mentioned, energy analysts see natural gas as a useful fuel to aid in the transition to a hydrogen economy, particularly because of its extensive distribution infrastructure. They do not see natural gas as a viable option for long-term large-scale hydrogen production because there is not enough of it. (If a way were found to extract the enormous deposits of methane hydrates that exist beneath the sea, the world's "energy equation" would be radically altered because methane would then become by far the largest source of fossil fuel energy in the world. But methane hydrates have never been produced at a commercial scale, and the first large-scale production wells lie at least several years in the future. It is not clear what contribution methane hydrates will make.)

Coal is, therefore, the only fossil fuel that is abundant enough and cheap enough to serve as a primary energy source and as a feedstock for the large-scale production of hydrogen.

The reasons, which are listed in chapter 7, that plant matter is currently ill-suited for use as a feedstock for a hydrogen economy are thought to preclude its use as a large-scale hydrogen source into the foreseeable future. In particular, there is just not enough farmland to produce the requisite amount of biomass and still leave sufficient capacity to produce the large amounts of high-quality affordable food to which people in most developed nations and many developing nations have become accustomed. And the production of hydrogen through genetically engineered microorganisms, though promising, remains in the earliest stages of development.

Water is, therefore, the only alternative to coal as a feedstock at the present time. Large-scale hydrogen production from water can occur in one of two ways: electrolysis, a process described in chapter 7, or thermochemistry. As mentioned previously, electrolysis is not an efficient way to produce hydrogen because it requires so much electricity. Hydrogen produced via electrolysis would be too expensive to be practical. Thermochemistry, which is also described in chapter 7, involves splitting water molecules (H_2O) into hydrogen and oxygen through the use of high-temperature chemical reactions. Specially designed high-temperature nuclear reactors are currently the only sources of energy capable of heating large amounts of water to the high temperatures required for this process to be successful on an industrial scale. This technology is many years away from commercialization.

There are several advantages to using nuclear reactors for large-scale hydrogen production. Their operation produces no greenhouse gas emissions; Western reactors have an excellent safety record; and obtaining nuclear fuel is much less disruptive than mining coal. Not as much uranium needs to be removed from the Earth as coal, and it is possible to design reactors that produce more fuel than they consume. Some have already been constructed.

Many experts believe that multiple satisfactory solutions to the problem of nuclear waste exist. Perhaps the most serious objection to the proliferation of nuclear reactors would involve the proliferation of dual-use technology, such as uranium enrichment technology, which can be used for peaceful purposes or to create the material needed to produce a nuclear bomb. But among those nations now in possession of nuclear weapons, Israel and Pakistan began their nuclear programs by manufacturing weapons and not by developing a commercial nuclear power program. Their behavior illustrates the point made by some that restricting the peaceful use of nuclear technology will not prevent nations that want nuclear weapons from developing them. The relationship between peaceful applications of nuclear power and nuclear weapons proliferation is complex and is an object of continued study and debate.

Whether large-scale coal or nuclear powered production facilities (or both) are used to generate sufficient hydrogen for a hydrogen

Oil pipeline, Kaparuk. A hydrogen economy would entail the construction of a pipeline system to rival the oil and natural gas pipeline systems already in place. *(William Breck Bowden)*

economy, the problem of bringing the hydrogen to market remains to be solved. Moving such large amounts of hydrogen requires an extensive pipeline network similar in concept to the one used to transport natural gas. Hydrogen has some special properties, however, that make a hydrogen pipeline system more expensive than the natural gas system; in particular, the natural gas pipeline system is ill-suited for the transport of hydrogen. An important characteristic of hydrogen is its density. Recall that hydrogen is much less dense than natural gas. One consequence of hydrogen's lower density is that in order to carry equivalent amounts of hydrogen at comparable pressures and temperatures, hydrogen pipelines must have diameters that are 50 percent larger than those used to transport natural gas. Hydrogen is also highly reactive and so it requires its own pipeline materials. If it were used in natural gas pipelines, the pipes and valves would eventually become brittle, increasing the possibility of breaks and leaks. Therefore, to transport hydrogen to market, a new pipeline network would have to be constructed. It is expected that the distribution cost for hydrogen may well prove to be the most expensive part of building a hydrogen economy. There is no easy or inexpensive way to implement a hydrogen-based economy.

Is the hydrogen economy worth the expense, both environmental and economic? Again, the answer depends on the alternatives. The most promising market for hydrogen is as a transportation fuel because eventually demand for oil will outstrip supply. The U.S. Energy Information Administration estimates that the global peak in oil production will occur between 2021 and 2067, with the most probable peak year being 2037. Until that time, the availability and price of oil will depend on a complex set of economic and political considerations. After peak oil has occurred, it will, as a matter of definition, be impossible to further increase the rate of oil production to meet what will presumably be the ever-increasing demand from the transportation sector for more fuel. After peak oil, many fuels that currently seem expensive or disadvantageous to use will begin to appear more attractive. Hydrogen may be one of those fuels.

Government Policies

In many nations, including the United States, a great deal of money is now spent researching hydrogen's potential as a transportation fuel. Additional money is spent investigating ways to use hydrogen for large-scale electricity generation. But such programs are not unique to hydrogen. The United States, for example, currently spends federal money researching coal, nuclear fission, solar, ethanol, biodiesel, nuclear fusion, wind, and many other sources or potential sources of usable energy. Governments have not always funded research into energy.

The oil business, for example, began with a period of explosive, subsidy-free growth. Its main product was initially kerosene, which was used as an illuminant. Compared to other liquid illuminants such as whale oil, which was more expensive, and tallow candles, which produced less light and more smoke per dollar, kerosene was a bargain right from the start. No subsidies were necessary. With

Los Angeles freeway. The scale of demand for transportation fuels makes fuel-switching particularly challenging. *(Joe Bellavia)*

the advent of the internal combustion engine, kerosene gave way to gasoline as the most profitable refinery product. Ethanol, a fuel whose use was originally supported by Henry Ford, was the main alternative to gasoline, but gasoline was cheaper than ethanol and released more thermal energy per unit volume. The differences between gasoline and ethanol were such that most people soon forgot about ethanol.

As the preceding chapters make clear, hydrogen is currently uncompetitive with other alternatives, and the fuels that hydrogen may one day replace remain superior to hydrogen in a number of ways. Few would even be considering hydrogen as a fuel except for two facts: First, oil supplies are often tight and prone to disruption, and second, there is a theoretical possibility that hydrogen can be manufactured at a reasonable cost and without the production of greenhouse gases. And so governments continue to fund hydrogen research programs.

Because hydrogen technology is currently so dependent on government support, it is important to understand the ways that hydrogen technology benefits from government support as well as the limitations of this type of support. This chapter describes connections between government policies and hydrogen-based technology development.

POLITICS VERSUS PHYSICS

Although hydrogen remains economically uncompetitive with fossil fuels and is likely to remain so for years to come, postponing research into hydrogen may not be wise. If research efforts are postponed until, for example, sufficient oil is unavailable at any price, widespread economic disruption and hardship may result when oil-dependent economies belatedly begin to search for ways of replacing an international trillion-dollar petroleum-dependent infrastructure. Consequently, a number of nations with strong technological and scientific resources have undertaken research into creating a hydrogen economy. (None of this is to say that a hydrogen economy is inevitable—only that it currently seems as promising as any of the other nonpetroleum alternatives.)

In the United States, government support for hydrogen research has been uneven. There was more support for hydrogen research during the years immediately following the 1973 OPEC oil embargo against the United States and the Netherlands, when oil prices were high, than there was in the late 1980s, when international oil prices collapsed and oil was once again a bargain. But oil prices did not stay down, and now with an understanding won through hard experience, an infrastructure for hydrogen research has been created in the United States, the European Union, and Japan that makes it more likely that hydrogen research will continue to be pursued independently of the price of oil. Taking into account that research was pursued unevenly in the past, what has been the result of hydrogen research so far?

Shortly after the 1973 embargo, under a grant from the National Science Foundation, a team of researchers wrote a long and thoughtful book assessing the possible advantages and disadvantages of a hydrogen economy. Entitled *The Hydrogen Energy Economy: A Realistic Appraisal of Prospects and Impacts,* this 1976 work lists the problems that would have to be overcome before a hydrogen economy could be implemented. In the mid-1970s the authors wrote

> ➢ Most hydrogen is produced from natural gas or naphtha, a liquid hydrocarbon. The authors do not even consider the possibility of producing hydrogen from natural gas over the long run because of insufficient supplies of natural gas. Water, they write, is a possible source of hydrogen, but it was, at the time, a minor source.

> ➢ With respect to the problem of storage of hydrogen as a transportation fuel, the authors assert that storing hydrogen as a gas in tanks is impractical because of high pressures and restrictions on the weight and size of tanks. They consider storage as a cryogenic liquid, but acknowledge the high cost of reducing hydrogen's temperature and the problem of the liquid boiling as the fuel warms. They discuss metal hydrides, a technology that involves storing hydrogen atoms in a hydrogen-metal lattice that would absorb and release hydrogen atoms in a way that is conceptually similar to the way a sponge absorbs and releases water, and they give a brief nod to the idea of storing hydrogen as a compound using the hydrogen-rich materials methanol and ammonia.

> ➢ At the time that the book was written, hydrogen deliveries were made by cryogenic tanker trucks and railcars.

They discuss the possibility of building a hydrogen pipeline network.

➤ Finally, the authors note that the absence of low-cost, high-efficiency catalysts remains a barrier to the widespread use of fuel cells.

Now contrast that mid-1970s list with some of the current barriers to the creation of a hydrogen economy

➤ Today, worldwide, 96 percent of all hydrogen is produced from fossil fuels, and only 4 percent of hydrogen is produced from water. In particular, natural gas remains the main source of hydrogen—almost 50 percent of all hydrogen is produced from natural gas.

➤ With respect to the storage of hydrogen, the choices for storage media are identical with those identified more than a generation ago. So are the challenges. There is still no way of storing a sufficient amount of hydrogen onboard a car to give it a cruising range similar to that of a gasoline-powered or diesel-powered automobile.

➤ Most hydrogen continues to be shipped by cryogenic trucks and railcars. A hydrogen pipeline network remains elusive. More than thirty years later, only several hundred miles of hydrogen pipeline have been built in the entire United States, slightly more in Europe.

➤ The 1976 remarks about the absence of low-cost, high-efficiency catalysts remain true today.

In fact, large sections of *The Hydrogen Economy* could be reproduced without change in a contemporary book about the hydrogen economy because not one of the main barriers to the creation of a hydrogen economy that was identified in the 1970s has been completely surmounted. This gives some indication of how difficult it is to solve the problems of production, transport, storage, and practical conversion of hydrogen to electricity via a fuel cell. Hydrogen, according to the authors of *The Hydrogen Economy*, could not contribute much to the United States' economy before the year

The Hydrogen Marketplace Today

A great many claims are made about the characteristics of a future hydrogen economy. To put these claims in perspective, it helps to know something of today's hydrogen market, and there is, in fact, a robust and substantial hydrogen market already in place. The hydrogen market is, for the most part, a captive market—that is, most hydrogen is produced where it is to be used by the company that plans to use it. (Hydrogen that is not part of the captive market is said to be part of the merchant market.) An example of a captive hydrogen market is the ammonia market. Many fertilizer factories, which are the main users of ammonia, produce hydrogen on-site and then use the hydrogen in the production of ammonia (chemical formula NH_3). (In the United States, almost 90 percent of ammonia is used in the production of fertilizer.)

Worldwide, roughly 55 million short tons (50 million metric tons) of hydrogen are consumed each year. Measured in terms of its energy content, this rate of consumption is somewhat less than 2 percent of all global energy consumption. The process by which this hydrogen is produced entails significant emissions: The production of 55 million short tons (50 million metric tons) of hydrogen releases about 350 million short tons (320 million metric tons) of CO_2 into the atmosphere. About half of all hydrogen consumed worldwide is used in the production of ammonia, and approximately 35 percent of the total is used by petroleum refineries. Methanol production takes 8 percent of the total. (Methanol is used in the production of formaldehyde, acetic acid, biodiesel, and a number of other products.)

In the United States, the hydrogen production figures are shifted more to the refinery industry, which produces approximately 3 billion

2000 and "even then only a few sectors would be affected." This very guarded forecast turned out to be too optimistic, but it was unusually pessimistic for the time. In another thoughtful book, *Hydrogen Technology for Energy*, also published in 1976, the authors identify

standard cubic feet (84 million m^3) of hydrogen each day, about half of all hydrogen produced in the United States. Approximately 35 percent of hydrogen production is used in the manufacturing of ammonia. The European hydrogen market is somewhat smaller than the one in the United States but displays the same general pattern of hydrogen consumption. By weight the United States produces about 9 million short tons (8 million metric tons) of hydrogen per year, a rate of production that requires about 5 percent of all the natural gas used in North American. Western Europe produces about 7 million short tons (6 million metric tons) per year. With respect to future growth, it is interesting to note that the fastest growing new market for hydrogen in North America is the oil sands project in Alberta, Canada. The oil sands are an enormous deposit of bitumen, a tar-like material that can be converted to an oil-like substance by a process that requires a lot of hydrogen. The hydrogen is produced by steam reforming of methane.

With ammonia and methanol production flat or decreasing, the short-term future of hydrogen in North America lies with the petroleum industry. Today, hydrogen is not an alternative transportation fuel but a material needed to manufacture high-quality gasoline. It is evident that hydrogen production and consumption do not have to be environmentally benign. If, from an environmental point of view, a hydrogen economy is to be an improvement over today's fossil fuel economy, it must be by design, because as today's consumption patterns show, there is nothing environmentally benign about the current hydrogen marketplace.

all of the same problems as the authors of *The Hydrogen Economy* but assert that the year 2000 is a reasonable date for a hydrogen economy to be in place.

This is the lesson: Despite more than a generation of on-again, off-again government research, not a single fundamental problem associated with the implementation of the hydrogen economy has been entirely overcome. What separates the hydrogen economy from certain other scientific research efforts—as, for example, missions to the Moon, computer-chip design, and the creation of the Internet—is that many of the barriers that have so far prevented the implementation of a hydrogen economy are closely associated with the *fundamental* properties of hydrogen, properties that can only be accommodated but not changed. This is the reason that political support, research, and subsidies have produced such modest results. Research programs can be legislated, but physics cannot.

CURRENT HYDROGEN RESEARCH

Creating a hydrogen economy is a series of challenges—some of which must be solved sequentially and some simultaneously, some of which are engineering or scientific in nature, and some of which are administrative or political. This volume has emphasized scientific and engineering aspects of those challenges because at today's level of development that is where most of the problems are to be found, but administrative and policy challenges are also important. Even when solutions to scientific and engineering problems are found, it does not, for example, follow that industry will undertake the necessary risks involved in attempting the transition to a hydrogen economy. Nor is there any guarantee that if offered the opportunity, consumers will embrace a hydrogen economy. Bridging the fossil fuel and hydrogen economies requires creative policies and skillful management. In the United States, risk aversion on the part of businesses is addressed when the government works to establish public-private partnerships, often international in nature, that encompass

California State Capitol. California has made great efforts in encouraging hydrogen-fueled vehicles, but market penetration of these vehicles is no better there than elsewhere. *(Kevin D. Korenthal)*

not just R & D (research and development) but R D & D (research, development, and demonstration).

The goal of a R D & D program is to shepherd a technology from abstract questions involving basic science to a working facility large enough to demonstrate both the technical and economic feasibility of the ideas in question. R D & D is, in a sense, an acknowledgement of the following very important fact: A scientific insight or laboratory device is not necessarily enough to attract investment. The reason is simple: Good ideas do not always scale up to good products.

The fact that private and government laboratories have worked to create a hydrogen economy since the 1970s is a good indication of how resistant to solution the problems of hydrogen production, storage, and consumption have so far proved to be. (Hydrogen research in the United States dates back to the Advanced Automotive Propulsion System Program of the 1970s, a program that predates the creation of the Department of Energy.) But as difficult as all of these engineering problems are, they are not equally difficult. After more than three decades of work, some are closer to solution than others. Like a group of hikers who begin hiking together in the morning only to find themselves scattered all along the path by the end of the day, some of the engineering problems associated with creating a hydrogen economy seem close to solution while others remain in the realm of basic research. As the technological complexity of creating a hydrogen economy has become more apparent, the government has sought to create new mechanisms, administrative and scientific, for moving forward.

Progress has not been smooth. As previously mentioned, FutureGen, a public-private partnership that attracted support from the China Huaneng Group, CONSOL Energy, Anglo American PLC, BHP Billiton, American Electric Power, and the government of India, among others, was abruptly cancelled in January 2008 by the U.S. Department of Energy, which cited cost overruns as the deciding factor. Having budgeted approximately $1 billion, the cost was later projected to be $1.8 billion. The 275-megawatt power plant would have been the world's first primarily hydrogen-fired power plant to employ carbon sequestration technology. The hydrogen would have been produced from coal. FutureGen was an opportunity to test all relevant technologies. As of 2009 the decision to cancel FutureGen is being reexamined.

The federal government is continuing with demonstration projects in carbon sequestration technology. Sequestration technology is, of course, not a necessary precondition for a hydrogen economy, but because coal seems to be the most likely source for large-scale quantities of hydrogen, hydrogen production without sequestration would have serious consequences for the world's climate.

The three Department of Energy–sponsored carbon sequestration demonstration projects—the Plains Carbon Dioxide Reduction Partnership, the Southeast Regional Carbon Sequestration Partnership, and the Southwest Regional Partnership for Carbon Sequestration—are being undertaken in partnership with neighboring states, and in the Plains Partnership, the Canadian provinces of Alberta, Saskatchewan, and Manitoba are also involved. These three projects average $100 million apiece with the costs shared among the federal, state, and provincial partners. The goal of each project is to store more than 1 million tons of carbon dioxide permanently beneath Earth's surface. The demonstrations will enable the participants to evaluate the entire process, including monitoring the sites after injection for signs that the carbon dioxide is moving or leaking. Evaluating the success of these projects is necessary in order to determine whether sequestration technology is sufficiently mature to deploy.

Within the Department of Energy, there is the Office of Fossil Energy, and within the Office of Fossil Energy, there is the Hydrogen from Coal Program. Hydrogen from Coal funds or conducts numerous smaller-scale programs, but nothing on the scale of FutureGen. Storing hydrogen, producing electricity from hydrogen, and other research projects designed to address those problems that have already been described in this book constitute many of its activities.

Specific issues related to the use of hydrogen as a transportation fuel are also investigated with the FreedomCAR initiative, a collaborative effort consisting of the U.S. Department of Energy, Chevron, ExxonMobil, Ford, General Motors, BP America, and other industry partners. This is a more nebulous program, however, and in particular, it is not solely devoted to creating a practical hydrogen-powered vehicle. The idea behind FreedomCAR is to develop technologies that reduce dependence on foreign oil and cut greenhouse gas emissions. Hydrogen technologies are included but so, for example, are programs aimed at reducing the cost and increasing the storage capacity of batteries. Other modestly-sized programs are to be found within other agencies.

If these federal programs seem small in scope relative to the problem of creating a hydrogen economy or if they seem haphazard in design, it is because they are. The United States government has yet to demonstrate a willingness to commit the funds or the scientific expertise necessary to effect a rapid transition to a hydrogen economy. Efforts are further hampered by a political system that sometimes distributes research dollars in ways that maximize political rather than engineering advantage. It is also true, however, that there is room for legitimate disagreement about the value of devoting enormous resources toward the creation of a hydrogen economy. A hydrogen economy is not, after all, an end in itself; it is a means to an end. The actual goal is to manufacture large amounts of power in a way that minimizes the environmental impact of manufacturing power. Not everyone believes that the creation of a hydrogen economy is the most efficient way of achieving that goal.

Finally, even if more resources could be brought to bear on the rapid transition to a hydrogen economy—and even if gross inefficiencies in resource allocation could be substantially reduced—a rapid transition to a hydrogen economy is by no means assured. The creation of a hydrogen economy is a very different kind of project than the large-scale engineering successes for which the United States is justifiably famous. Earlier large-scale government engineering efforts—the Manhattan Project of World War II, for example, or the Apollo missions—were technically very difficult, but they were much narrower in scope. The Manhattan Project was completed when the first nuclear bomb successfully detonated in New Mexico, and it is not too much of an exaggeration to say that the Apollo project was completed when the first astronauts returned from walking on the Moon. The attempt to create a hydrogen economy has already taken longer than either the Apollo or

Manhattan projects, and it will have a more ambiguous completion date. But if the effort is successful, it will transform the world. It will profoundly change many economic, political, and military relationships among nations. *If* successful, the efforts of the engineers and scientists involved will directly and permanently impact the lives of each individual on Earth. It will not only affect how people travel; it will also change the environmental implications of travel. The same can be said of its effects on the electric power industry. Even the evolution of the global climate would be altered from its present course. The program to create a hydrogen economy is one of the most ambitious projects ever undertaken, and as with all good research projects, success is not guaranteed.

# Chronology

ca. 221 C.E.	History's first gas wells are dug in China. The gas is transported via a bamboo pipeline network.
1792	British inventor William Murdock (1754–1839) demonstrates coal gas technology by lighting his own home.
1811	British physicist Humphrey Davy publishes a paper describing the first clathrate hydrate to be discovered. It is a material that is similar in structure to methane hydrate, but composed of water and chlorine.
1812	London & Westminster Gas Light & Coke Company, the world's first coal gas company, begins operation.
1820	William Hart of Fredonia, New York, a gunsmith and local entrepreneur, drills for gas and strikes it at a depth of 27 feet (8 m). The gas is piped to a local inn. It is the first commercial gas well since those of the Chinese.
1839	British physicist and jurist William Grove (1811–96) discovers the principle of the fuel cell.
1858	The first natural gas company, the Fredonia Gas Light Company, is formed in New York.
1859	There are 297 coal gas companies in the United States supplying 4,857,000 customers.
1870	The Bloomfield & Rochester Natural Gas Light Company is formed to transport natural gas from West Bloomfield, New

York, to Rochester, New York, via pipeline. The ambitious project collapses soon after completion of the pipeline.

Also about this time a number of oil drillers discover the Trenton Field of Indiana. It was, at the time, the world's largest known natural gas field.

1872   The New York Mercantile Exchange (NYMEX), pioneering institution in the creation of energy futures markets, is founded.

1885   The Austrian engineer and chemist Carl Auer, baron von Welsbach (1858–1929) patents the Welsbach light, which enabled gaslights to burn much more brightly than a conventional gas light, thereby prolonging the life of the gaslight industry.

1891   The Indiana Natural Gas and Oil Company installs twin eight-inch natural gas pipelines a distance of 120 miles to connect Chicago with the gas fields to the south.

1902   Most gas wells in the Trenton Field have by this time stopped producing, reflecting the fact that an enormous natural resource was largely wasted by unregulated drilling and inefficient consumption.

1920   Congress established the Federal Power Commission, the forerunner of today's Federal Energy Regulatory Commission, and the first federal agency to regulate the natural gas business.

1938   Congress passes the Natural Gas Act, the first federal legislation regulating the interstate transport of natural gas.

1954   In *Phillips Petroleum v. Wisconsin,* the United States Supreme Court ruled that in addition to interstate pipeline operators, natural gas producers who sold gas that was transported by interstate pipelines were also subject to regulation.

1965 NASA's Gemini 5 space mission launched. It is NASA's first test of a fuel cell in space. The fuel cell failed.

1971 Research Applied to National Needs (RANN), a farsighted federal program that attempts to harness scientific research to serve society, is founded. Among its programs is the Advanced Automotive Propulsion System Program, which investigates using hydrogen as a transportation fuel.

During the 1970s natural gas shortages begin to develop due to inept federal regulation of the natural gas business.

1977 The United States Department of Energy is established.

1978 In response to natural gas shortages, Congress passes the Natural Gas Policy Act of 1978 (NGPA), which grants the Federal Energy Regulatory Commission authority over intrastate as well as interstate production.

1985 The Federal Energy Regulatory Commission issues Order 436, marking the beginning of the modern natural gas markets in the United States.

1990 Congress passes the Spark M. Matsunaga Hydrogen Research, Development, and Demonstration Program Act for the purpose of accelerating efforts to "economically produce hydrogen in quantities that will make a significant contribution toward reducing the Nation's dependence on conventional fuels."

NYMEX begins trading Henry Hub natural gas futures contracts.

1992 Congress passes the Energy Policy Act of 1992, which directs the Secretary of the U.S. Department of Energy to implement a program of "research, development, and demonstration of fuel cells and related systems for transportation applications. . . ."

FERC extends Order 436 by issuing Order 636, which creates the restructured natural gas markets in essentially the form in which they exist today.

1997   Natural gas futures begin to trade on the International Petroleum Exchange, which is located in London.

2000   Congress passes the Methane Hydrate Research and Development Act to speed the development of this vast resource.

2003   In his State of the Union Address, President Bush announces a $1.2 billion Hydrogen Fuel Initiative with the goal of making hydrogen a competitive transportation fuel by 2020.

2005   The Energy Policy Act of 2005 is passed. It directs the Secretary of the U.S. Department of Energy to conduct R & D in "technologies related to the production, purification, distribution, storage, and use of hydrogen energy, fuel cells, and related infrastructure."

2007   The Department of Energy, the U.S. Geological Survey, and BP undertake a project on Alaska's North Slope to demonstrate the commercial production of methane hydrates.

2008   FutureGen is canceled by the Department of Energy.

List of Acronyms

Btu British thermal unit

CCGT combined cycle gas turbine

EIA Energy Information Administration

FERC Federal Energy Regulatory Commission

HHV higher heating value

IGCC Integrated Gasification Combined Cycle

LHV lower heating value

LNG liquefied natural gas

NGPA Natural Gas Policy Act

OPEC Organization of Petroleum Exporting Countries

PEM proton exchange membrane

R & D research and development

R D & D research, development, and demonstration

 # Glossary

anticline an arch-like or dome-like formation of rock strata that can act as a petroleum trap

biomass plant or animal material used as fuel or **feedstock**

carbon sequestration the storage of carbon dioxide in a geological formation or the fixation of carbon in the ocean or in living matter (as in a forest)

cash market the **commodity** market in which the physical product is traded

catalyst a material that enables a chemical reaction to take place at a faster rate than would occur without it

coal gas an early gaseous fuel produced by heating coal in a low-oxygen environment

commodity an unspecialized bulk product produced for sale

density the ratio formed by the mass of a sample of material to its volume

electrolysis the process by which hydrogen is produced by passing an electric current through water

feedstock the raw material supplied for a manufacturing process

fuel cell a device created to convert hydrogen and oxygen directly into electricity

futures contract a standardized contract for the purchase or sale of a **commodity**

futures market an open market for the trading of **futures contracts**

gas cap a gas deposit that forms beneath an **anticline** and above an oil deposit

gasworks a facility in which coal gas is produced

generator a device that converts rotational energy into electrical energy

greenhouse gas a gas that increases the heat-retention properties of the atmosphere as its atmospheric concentration increases

heat exchanger a device that allows two fluids to exchange thermal energy while keeping them physically separate

hedge a form of financial protection used to protect a market participant against fluctuations in the price of a **commodity**

higher heating value (HHV) the measure of the energy released by a combustion reaction when the water vapor produced by the reaction is allowed to condense before the measurement of the amount of thermal energy released is made

hydrate stability zone those areas on Earth where the pressure and temperature conditions are conducive to the formation of **methane hydrates**

hydraulic fracturing the process by which water is injected into a geological formation to create cracks through which petroleum can flow

hydrogen economy an economy in which a major fossil fuel, usually petroleum, is replaced by hydrogen

lower heating value (LHV) the measure of the energy released by a combustion reaction when the measurement of the amount of thermal energy released by the reaction is made before the water present in the combustion gases condenses

market participant one who participates in a financial or **cash market**

methane hydrate a crystalline structure formed by methane and water molecules

metric ton a unit of measurement equaling 1,000 kilograms

reformer a device for converting methane into hydrogen

reservoir rock a porous rock that contains petroleum

short ton a unit of measurement equaling 2,000 pounds

speculator a **market participant** who attempts to profit by anticipating price fluctuations in an underlying **commodity**

standard cubic foot the quantity of gas occupying one cubic foot (2.83 l) of space at one atmosphere pressure at 60°F (15.6°C)

tight gas   gas in a geological formation that is not conducive to the flow of gas

trader   a market participant whose business is buying and selling the commodity of interest

turbine   a device for converting the linear or straight-line motion of a liquid or gas into rotary motion

 **Further Resources**

Both natural gas and hydrogen have interesting histories as well as potentially interesting futures. In fact, because of the vast quantities of methane hydrates that have already been identified, either natural gas or hydrogen may prove to be "the fuel of the future." The following list of resources provide good starting places for those interested in the past, present, and future of these important fuels.

BOOKS

Dickson, Edward M., John W. Ryan, and Marilyn H. Smulyan. *The Hydrogen Energy Economy: A Realistic Appraisal of Prospects and Impacts.* New York: Praeger Publishers, 1977. This is a well-written book. Because it was published more than 30 years ago, it should be out of date. It is not. The barriers to the creation of a hydrogen economy that were described 30 years ago remain the barriers of today. Reading this book gives a sense of perspective that one cannot gain by reading the many accounts, old and new, that fail to appreciate just how difficult it is to overcome the engineering barriers that continue to prevent the creation of a hydrogen economy.

Errera, Steven, and Stewart L. Brown. *Fundamentals of Trading Energy Futures and Options.* 2nd ed. Tulsa: PennWell Corporation, 2002. For those interested in the relationships that exist between physical and financial markets, this is a good introduction. The presentation involves no advanced mathematics, and the text is accompanied by a good glossary to help the reader decipher the financial jargon, which is both dense and unavoidable.

National Academy of Engineering. *The Carbon Dioxide Dilemma: Proceedings of a Symposium, April 23–24, 2002.* Washington, D.C.: National Academies Press, 2003. This publication consists of a series of not-especially-technical papers by engineers, physicists, climatologists, economists, oceanographers, and others about global climate change and strategies to control it. This book can also be read on the Web, albeit in a very non-user-friendly format. URL: http://books.nap.edu/openbook. php?record_107988page=R10. Accessed on September 1, 2007.

National Research Council and National Academy of Engineering. *The Hydrogen Economy: Opportunities, Costs, Barriers, and R & D Needs.* Washington, D.C.: National Academies Press, 2004. This is a cautious and lucid introduction to current thinking about the hydrogen economy.

National Research Council. *Summary of a Workshop on U.S. Natural Gas Demand, Supply, and Technology: Looking toward the Future.* Washington, D.C.: National Academies Press, 2003. A thoughtful discussion of the growing gap between domestic demand and supply and the ways that this gap might be bridged.

Pool, Robert. *Beyond Engineering: How Society Shapes Technology.* New York: Oxford University Press, 1997. An interesting book that is concerned with the general problem of how society affects technology. While not specifically about the hydrogen economy, it offers general insight into the way that a hydrogen economy has and has not developed.

Schlenker, Sandra Noreen Bortle. *Windows to the Past.* Interlaken, N.Y.: Heart of the Lakes, 2003. A good source of information about some of the earliest days of the natural gas business in the area in and around West Bloomfield, New York. It is rich in original sources.

Stotz, Louis, and Alexander Jamison. *History of the Gas Industry.*
New York: Stettiner Brothers, 1938. This is a history of the coal
gas industry, and it contains a remarkable number of interesting
stories and unusual facts.

Tabak, John. *Mathematics and the Laws of Nature: Developing the
Language of Science.* New York: Facts On File, 2004. There are
severe limitations on the efficiency of heat engines. Understanding what these are is important if one is to appreciate conventional power plant technology and its limitations. (See chapter 7
of this book for a condensed discussion of this vital topic.)

———. *Coal and Oil.* New York: Facts On File, 2009. Understanding the hydrogen economy means understanding coal and
oil—coal, because it will probably be the principal feedstock for
a future hydrogen economy, and oil, because it is oil with which
hydrogen must compete as a transportation fuel.

INTERNET RESOURCES

Energy Information Administration. "Natural Gas Navigator:
Analysis for Natural Gas Basics." Available online. URL: http://
tonto.eia.doe.gov/dnav/ng/ng_pub_analysis_basics.asp. Accessed on October 16, 2007. The EIA's site, which is not especially easy to navigate, is the most comprehensive online source
of information about the natural gas business. The articles range
from elementary to technical. This particular page contains an
excellent overview of many important aspects of the natural
gas business. These articles are accessible to any reader who has
completed this volume.

Energy Information Administration. "Natural Gas: Major Legislative and Regulatory Actions (1938–2004)." Available online.
URL: http://www.eia.doe.gov/oil_gas/natural_gas/analysis_
publications/ngmajorleg/ngmajorleg.html. Accessed on April
24, 2008. An excellent history that illustrates the advantages
and disadvantages of government regulation of the marketplace.

Lerner, Robert L., and Timothy J. Rudderow. "The Mechanics of the Commodity Futures Markets: What They Are and How They Function." Available online. URL: http://www.turtletrader. com/beginners_report.pdf. Accessed on October 16, 2007. A brief, well-written introduction to the commodity futures markets.

Natural Gas Supply Association. NaturalGas.org. Available online. URL: http://www.naturalgas.org. Accessed on October 16, 2007. This site contains a remarkably complete and thoughtful presentation of many of the ideas and technologies that underpin the natural gas business. The articles are nontechnical and provide an excellent and very basic introduction to the subject.

United States Department of Energy. Fossil Energy Techline. "DOE Awards First Three Large-Scale Carbon Sequestration Projects." Available online. URL http://www.fossil.energy. gov/news/techlines/2007/07072-DOE_Awards_Sequestration_ Projects.html. Accessed on October 16, 2007. A brief report on the state of the art in carbon sequestration.

———, Office of Fossil Energy and the National Energy Technology Laboratory. "Hydrogen from Coal Program: Research, Development, and Demonstration Plan for the Period 2007 through 2016." Available online. URL: http://www.fossil.energy. gov/programs/fuels/publications/programplans/2007_Hydro- gen_Program_Plan.pdf.   Accessed on October 16, 2007. This 80-page report provides a good overview of the current state of research into the technology necessary to create a hydrogen economy.

United States Geological Survey. "Gas Hydrate Studies—A Part of the Geophysics Group." Available online. URL: http://woods hole.er.usgs.gov/project-pages/hydrates/. Accessed on April 24, 2008. This page, together with its links, provides a quick overview of methane hydrate formations and their potential as an energy source.

 **Index**

W